Japanese Telecommunications

Using previously unprinted material, and material printed in English for the first time, this book presents a comprehensive survey of the telecommunications industry in Japan. It covers the different sectors of the industry – including mobile, broadband and satellite – and considers key questions such as the structure and economics of the industry, government policy, and international relations issues connected to the industry.

The volume brings together unique analysis by renowned experts in the telecommunications field. One major overall problem is that, unlike in many other industries, Japan once lagged behind other countries in telecommunications. This is considered to be due to the monopolistic industry structure and policy taken in the past, which this book explores in detail.

After over a decade of struggle, Japan has recorded rapid uptake of broadband, and Japanese advanced mobile services have become increasingly successful on a global scale. Japan has also undergone regulatory reform, and competition policy is now given top priority by government. This book examines the most recent developments and provides signposts for the future.

Ruth Taplin is the Director of the Centre for Japanese and East Asian Studies. The Centre won Exporter of the Year in Partnership in Trading/Pathfinder for the United Kingdom in 2000. She received her doctorate from the London School of Economics and is the author/editor of 11 books, the most recent of which is *Risk Management and Innovation in Japan, Britain, and the United States* (Routledge, 2005). She has been editor of the *Journal of Interdisciplinary Economics* for ten years.
Masako Wakui is an Associate Professor at the Graduate School of Law, Osaka City University, Japan. She has been conducting research on the information communication industry and IPR-related matters from competition policy perspectives and has published numerous book and journal articles.

Routledge studies in the growth economies of Asia

1 **The Changing Capital Markets of East Asia**
Edited by Ky Cao

2 **Financial Reform in China**
Edited by On Kit Tam

3 **Women and Industrialization in Asia**
Edited by Susan Horton

4 **Japan's Trade Policy**
Action or reaction?
Yumiko Mikanagi

5 **The Japanese Election System**
Three analytical perspectives
Junichiro Wada

6 **The Economics of the Latecomers**
Catching-up, technology transfer and institutions in Germany, Japan and South Korea
Jang-Sup Shin

7 **Industrialization in Malaysia**
Import substitution and infant industry performance
Rokiah Alavi

8 **Economic Development in Twentieth Century East Asia**
The international context
Edited by Aiko Ikeo

9 **The Politics of Economic Development in Indonesia**
Contending perspectives
Edited by Ian Chalmers and Vedi Hadiz

10 **Studies in the Economic History of the Pacific Rim**
Edited by Sally M. Miller, A. J. H. Latham and Dennis O. Flynn

11 **Workers and the State in New Order Indonesia**
Vedi R. Hadiz

12 **The Japanese Foreign Exchange Market**
Beate Reszat

13 **Exchange Rate Policies in Emerging Asian Countries**
Edited by Stefan Collignon, Jean Pisani-Ferry and Yung Chul Park

14 **Chinese Firms and Technology in the Reform Era**
Yizheng Shi

15 **Japanese Views on Economic Development**
Diverse paths to the market
Kenichi Ohno and Izumi Ohno

16 **Technological Capabilities and Export Success in Asia**
Edited by Dieter Ernst, Tom Ganiatsos and Lynn Mytelka

17 **Trade and Investment in China**
The European experience
Edited by Roger Strange, Jim Slater and Limin Wang

18 **Technology and Innovation in Japan**
Policy and management for the 21st century
Edited by Martin Hemmert and Christian Oberländer

19 **Trade Policy Issues in Asian Development**
Prema-chandra Athukorala

20 **Economic Integration in the Asia Pacific Region**
Ippei Yamazawa

21 **Japan's War Economy**
Edited by Erich Pauer

22 **Industrial Technology Development in Malaysia**
Industries and firm studies
Edited by K. S. Jomo, Greg Felker and Rajah Rasiah

23 **Technology, Competitiveness and the State**
Malaysia's industrial technology policies
Edited by K. S. Jomo and Greg Felker

24 **Corporatism and Korean Capitalism**
Edited by Dennis L. McNamara

25 **Japanese Science**
Samuel Coleman

26 **Capital and Labour in Japan**
The functions of two factor markets
Toshiaki Tachibanaki and Atsuhiro Taki

27 **Asia Pacific Dynamism 1550–2000**
Edited by A. J. H. Latham and Heita Kawakatsu

28 **The Political Economy of Development and Environment in Korea**
Jae-Yong Chung and Richard J. Kirkby

29 **Japanese Economics and Economists since 1945**
Edited by Aiko Ikeo

30 **China's Entry into the World Trade Organisation**
Edited by Peter Drysdale and Ligang Song

31 **Hong Kong as an International Financial Centre**
Emergence and development 1945–1965
Catherine R. Schenk

32 **Impediments to Trade in Services**
Measurement and policy implication
Edited by Christoper Findlay and Tony Warren

33 **The Japanese Industrial Economy**
Late development and cultural causation
Ian Inkster

34 **China and the Long March to Global Trade**
The accession of China to the World Trade Organization
Edited by Alan S. Alexandroff, Sylvia Ostry and Rafael Gomez

35 **Capitalist Development and Economism in East Asia**
The rise of Hong Kong, Singapore, Taiwan, and South Korea
Kui-Wai Li

36 **Women and Work in Globalizing Asia**
Edited by Dong-Sook S. Gills and Nicola Piper

37 **Financial Markets and Policies in East Asia**
Gordon de Brouwer

38 **Developmentalism and Dependency in Southeast Asia**
The case of the automotive industry
Jason P. Abbott

39 **Law and Labour Market Regulation in East Asia**
Edited by Sean Cooney, Tim Lindsey, Richard Mitchell and Ying Zhu

40 **The Economy of the Philippines**
Elites, inequalities and economic restructuring
Peter Krinks

41 **China's Third Economic Transformation**
The rise of the private economy
Edited by Ross Garnaut and Ligang Song

42 **The Vietnamese Economy**
Awakening the dormant dragon
Edited by Binh Tran-Nam and Chi Do Pham

43 **Restructuring Korea Inc.**
Jang-Sup Shin and Ha-Joon Chang

44 **Development and Structural Change in the Asia-Pacific**
Globalising miracles or end of a model?
Edited by Martin Andersson and Christer Gunnarsson

45 **State Collaboration and Development Strategies in China**
The case of the China–Singapore Suzhou Industrial Park (1992–2002)
Alexius Pereira

46 **Capital and Knowledge in Asia**
Changing power relations
Edited by Heidi Dahles and Otto van den Muijzenberg

47 **Southeast Asian Paper Tigers?**
From miracle to debacle and beyond
Edited by Jomo K.S.

48 **Manufacturing Competitiveness in Asia**
How internationally competitive national firms and industries developed in East Asia
Edited by Jomo K.S.

49 **The Korean Economy at the Crossroads**
Edited by MoonJoong Tcha and Chung-Sok Suh

50 **Ethnic Business**
Chinese capitalism in Southeast Asia
Edited by Jomo K.S. and Brian C. Folk

51 **Exchange Rate Regimes in East Asia**
Edited by Gordon De Brouwer and Masahiro Kawai

52 **Financial Governance in East Asia**
Policy dialogue, surveillance and cooperation
Edited by Gordon De Brouwer and Yunjong Wang

53 **Designing Financial Systems in East Asia and Japan**
Edited by Joseph P. H. Fan, Masaharu Hanazaki and Juro Teranishi

54 **State Competence and Economic Growth in Japan**
Yoshiro Miwa

55 **Understanding Japanese Saving**
Does population aging matter?
Robert Dekle

56 **The Rise and Fall of the East Asian Growth System, 1951–2000**
International competitiveness and rapid economic growth
Xiaoming Huang

57 **Service Industries and Asia-Pacific Cities**
New development trajectories
Edited by P. W. Daniels, K. C. Ho and T. A. Hutton

58 **Unemployment in Asia**
Edited by John Benson and Ying Zhu

59 **Risk Management and Innovation in Japan, Britain and the USA**
Edited by Ruth Taplin

60 **Japan's Development Aid to China**
The long-running foreign policy of engagement
Tsukasa Takamine

61 **Chinese Capitalism and the Modernist Vision**
Satyananda J. Gabriel

62 **Japanese Telecommunications**
Market and policy in transition
Edited by Ruth Taplin and Masako Wakui

Japanese Telecommunications

Market and policy in transition

Edited by Ruth Taplin and Masako Wakui

Routledge
Taylor & Francis Group
LONDON AND NEW YORK

First published 2006
by Routledge
2 Park Square, Milton Park, Abingdon, Oxon OX14 4RN

Simultaneously published in the USA and Canada
by Routledge
270 Madison Ave, New York, NY 10016

Routledge is an imprint of the Taylor & Francis Group

Typeset in Times by Wearset Ltd, Boldon, Tyne and Wear
Printed and bound in Great Britain by MPG Books Ltd, Bodmin

British Library Cataloguing in Publication Data
A catalogue record for this book is available from the British Library

Library of Congress Cataloging in Publication Data
A catalog record for this book has been requested

ISBN 0-415-36803-0

Contents

Figures

Tables

Contributors

Sumiko Asai is an Associate Professor at the School of Social Information Studies, Otsuma Women's University, Japan. Her recent work includes *Integration and Modularity in the Information Industry* (in Japanese) (Tokyo: Nippon Hyoron, 2004); *The Policy Evaluation of Information Communication Policy* (in Japanese) (Tokyo: Nihon Hyoron, 2001) and *Economic Analysis of Telecommunication Business* (in Japanese) (Tokyo: Nihon Hyoron, 1999); and 'Factor Analysis of Demand Growth for Information Technology Input in Japan', *Economics of Innovation and New Technology* 13 (8): 687–694 (2004).

Jeffrey L. Funk is Professor of Business at Hitotsubashi University's Institute for Innovation Research. He received his Ph.D. from Carnegie Mellon University in 1984 in the Department of Engineering and Public Policy. His research primarily focuses on managerial and economic issues associated with innovation, and his recent books include *Global Competition between and within Standards: The case of mobile phones* (London: Palgrave, 2002) and *Mobile Disruption: The technologies and applications driving the mobile Internet* (New York: John Wiley, 2004). He was awarded the 2004 DoCoMo Mobile Science Award for lifetime contributions to the field of mobile communications.

Takanori Ida is an Associate Professor at the Graduate School of Economics, Kyoto University. His interest is in network economics and he studied the industry both empirically and theoretically, which resulted in *Network Economics* (in Japanese) (Tokyo: Nihon-Hyoron kk., 2001); 'The Interconnection and Pricing of the Internet' (with M. Ueda), in M. Kagami, M. Tsuji and E. Giovannetti (eds) *Information Policy and the Digital Divide* (Cheltenham and Northampton, UK: Edward Elgar, 2004, pp. 282–302); 'The Cost Structure of the Japanese Railway Industry: The Economies of Scale and Scope and the Regional Gap of the Japan Railway after the Privatization' (with M, Suda), *International Journal of Transport Economics* 31 (1): 23–37 (2004); 'Regulatory Reform of Japan's Electric Power Industry: Economies of Scale-and-Scope and Yardstick Competition' (with T. Kuwahara), *Asian Eco-*

nomic Journal 18 (4): 423–438 (2004); and 'Price Discrimination and Universal Service', *The Kyoto University Economic Review* 150/151: 43–64 (2002).

Hajime Oniki is a Professor at Osaka-Gakuin University and Professor Emeritus of Osaka University, Japan. He is scholar in economics and his research has been wide-ranging, including telecommunication policies. On spectrum policy, his analyses and views have been presented at international conferences, in journals and governmental workshops, such as 'Allocation and Assignment of Radio-Spectrum Resources by Using Price Mechanism: Proposals for a New System', presented as a plenary speech at the Workshop on Advanced Wireless Technologies: Implications for Spectrum Management, European Commission, DG Information Society, IST Program, on 10 October 2003. In Japan, through his book *Economics of Radiowave Spectrum: A study on spectrum auction in the U.S.* (in Japanese) (Kanagawa: Gendai Tosho, 2002) and as a member of government spectrum committees, he has been working to improve spectrum policy in Japan. He has worked for several professional societies as an editor: *International Economic Review* (Editor, 1986–1990; Associate Editor, 1990–1996); *Information Economics and Policy* (Associate Editor, 1984–2002); *Economics of Innovation and New Technology* (Member of Editorial Board, 1989–2000); *info* (Member of Editorial Board, 1998 to date); and *Telecommunications Policy* (Member of International Editorial Board, 1999 to date).

Kenji Suzuki is an Associate Professor, European Institute of Japanese Studies, Stockholm School of Economics, Sweden. He is the author of the book *Competition Law Reform in Britain and Japan: Comparative analysis of policy network* (Routledge 2002). His studies have ranged over governmental–industrial relationships in Japan and also public policy, policy-making processes and public management in Japan and European countries. His work has been published in 'The Effect of Amakudari on Bank Performance in the Post-bubble Period', in Sung-Jo Park and Sierk Horn (eds) *Asia and Europe in the New Global System* (London: Palgrave Macmillan, 2003) and as working papers of the European Institute of Japanese Studies, Stockholm School of Economics: 'Organizational Learning Capacity of Policy-Making System for Financial System Recovery: The Case of Sweden with some Suggestions to Japan', Working Paper No. 148 (2002); 'Competition Policy in the Japanese Banking Sector: Support Big Bang?', Working Paper No. 113 (2001); and 'Changing the Swedish Welfare State: Towards a More Flexible Model' (in Japanese), Working Paper No. 112 (2001).

Ruth Taplin is the Director of the Centre for Japanese and East Asian Studies. The centre won Exporter of the Year in Partnership in

Trading/Pathfinder for the UK in 2000. She received her doctorate from the London School of Economics and is the author/editor of 11 books. The most recent are an edited series, *Exploiting Patent Rights and a New Climate for Innovation in Japan* (London: Intellectual Property Institute, 2003), *Valuing Intellectual Property in Japan, Britain and the United States* (London: RoutledgeCurzon, 2004) and *Risk Management and Innovation in Japan, Britain, and the United States* (London: RoutledgeCurzon, 2005). Professor Taplin has been Editor of the *Journal of Interdisciplinary Economics* for ten years, is the author of numerous articles and is on the editorial board of *Managerial Accounting Journal*. She has been on television and radio around the world speaking on topics within economics, IP, law and cross-cultural matters in relation to East Asia. She has been a consultant to the Federation of the Electronics Industry (Intellect) for ten years, bringing speakers to the Asia Pacific, dealing with IP and GPS matters. Currently she is a Research Fellow at Birkbeck College, University of London, and the University of Leicester. Recently, she was appointed Visiting Professor at the School for International Business and Management at the University of Warsaw, Poland.

Motohiro Tsuchiya is an Associate Professor at Graduate School of Media and Governance, Keio University, Japan. Dr Tsuchiya is interested in the impact of information technologies on global governance. He authored *Information and Global Governance* (in Japanese) (Tokyo: Keio University Press, 2001), *Net Politics* (in Japanese) (Tokyo: Iwanami Publishing, 2003), which was awarded the Telecommunications Advancement Foundation Social Science Prize, and many other books and articles. He was a Visiting Scholar at the University of Maryland and George Washington University from 2001 to 2002.

Masako Wakui is an Associate Professor at the Graduate School of Law, Osaka City University, Japan. She has been conducting research on the information communication industry and IPRs-related matters from competition policy perspectives, which has resulted in works such as 'Standardisation and Patent Pools in Japan', in Ruth Taplin (ed.) *Valuing Intellectual Property in Japan, Britain and the United States* (London: RoutledgeCurzon, 2004); *Economic Law Basics* (Tokyo: Yuhikaku, 2003) (co-author; in Japanese); 'Standard setting and Antimonopoly Law', *Hokkaido Law Review* 53: 1048–1103 (2002) (in Japanese). She visited the London School of Economics and Political Science as a Nitobe Fellow, International House of Japan (2002–2003) and the Institute for Computer and Communications Law, Queen Mary College, University of London (2003–2004).

Preface

This book brings to English-language readers a unique analysis by renowned experts in the telecommunications field. Recent changes in the telecommunications sector has been a real catalyst to improving the state of the Japanese economy. The lag of advanced communication technology, especially the Internet, was said to be largely responsible for the 'Japanese lost decade'. or the serious recession of the Japanese economy. Onerous regulations and the lack of a strong competition policy were criticized as the basis of such failure. Not content with an antiquated regulatory process lacking in transparency, foreign companies grew critical of the Japanese regulatory system. In the case of the United States, questions have been asked concerning its role in driving Japanese reform. Telecommunications has given rise to diplomatic issues that can have serious implications for the geopolitical global balance of power.

After over a decade of struggle, Japan has recorded rapid uptake of broadband. Japanese advanced mobile services have become increasingly successful on a global scale. Japan has also undergone regulatory reform. Competition policy is now given top priority in government official policy, and various regulations to promote competition have been introduced.

This book focuses on essential issues that define the current Japanese market and regulations, which are analysed within an internationally grounded framework and conceptual structures. The book is objective in the sense that the analysis is based on verifiable and empirical facts as much as possible. These views, however, are analytical as they are not just a description of the current state of affairs. Readers may know basic facts about markets and regulations in Japan, but through this book may gain new insights concerning telecommunications policy applicable in many countries and global regulations from an economic, institutional and geopolitical standpoint. Recommendations in this book will have direct relevance to the global telecommunications order.

Acknowledgements

The editors would like to thank Professor Akira Negishi of Kobe University, Japan, for his kind support, which was greatly appreciated, coming from the very best competition and regulation policy scholar in Japan. His encouragement has been a great impetus for this project.

Ruth Taplin would like to thank the Daiwa Anglo-Japanese Foundation, for the grant given towards her research trip to Japan in April 2004. She would also like to thank Mr Hiroshi Nishiguchi, Secretary-General of the Japan GPS Council for his substantive contribution to the satellite chapter and Professor Terry Moore, Professor of Satellite Navigation and Director of the Institute of Engineering Surveying and Space Geodesy, University of Nottingham for his insightful comments and contributions, Mr F. Michael Swiek, Executive Director, United States GPS Industry Council, Mr Akito Tani, formerly Deputy General JETRO London, Mr Yutaka Harada, Senior Adviser to JETRO London and the Soumusho for their assistance with research information.

Masako Wakui would like to thank Dr Ian Walden, Head of the Institute for Computer and Communications Law, Queen Mary College, University of London, and his colleagues. The Institute proved to be an excellent place for telecommunications law study, where she was always provided with essential, current information, expert advice and a pleasant study environment.

She is most appreciative of support given by the Telecommunications Advancement Foundation, Japan and International House of Japan during her stay in London. There, she met many telecommunications experts who devote themselves to the pursuit of improved markets and regulations. Encouraged by them, she took on the challenge of this book project.

Abbreviations

ADSL	asymmetrical digital subscriber line
ARIB	Association for Radio Industry Businesses
ASBC	Advanced Space Business Corporation
BIPM	Bureau International des Poids et Mesures
B-ISDN	broadband integrated services digital network
BS	broadcasting satellite
C/C	command and control
CATV	cable television
CCITT	International Telegraph and Telephone Consultative Committee
CCTF	Consultative Committee for Time and Frequency
cdma	code division multiple access
CEO	chief executive officer
CGGTTS	CCTF Group for GPS Time Transfer Standardization
c-HTML	compact HTML
CPE	customer premises equipment
CRL	Communications Research Laboratory
CS	communications satellite
DGPS	Differential Global Positioning System
DP	Democratic Party
DTV	digital television
DU	dial-up
ETSI	European Telecommunications Standards Institute
EU	European Union
FCC	Federal Communications Commission [United States]
FDMA	frequency division multiple access
FTC	Fair Trade Commission
FTTC	fibre to the curb
FTTH	fibre to the home
FWA	fixed wireless access
FY	fiscal year
GATT	General Agreement on Tariffs and Trade
GBAS	Ground-Based Augmentation System

GEONET	GPS Earth Observation Network
GGPS	Galileo Global Positioning System
GIS	Geographic Information System(s)
GNSS	Global Navigation Satellite Service
GPS	Global Positioning System(s)
GSI	Geographical Survey Institute
GSM	Global System for Mobile Communications
HD	high definition
HDD	hard disk drive
HDTV	high-definition television
HELPNET	Help systems for Emergency Life saving and Public safety Network
HHI	Hirfindahl–Hirshman Index
HTML	Hypertext Mark-up Language
ICT	information and communications technology
IDO	Nippon Idou Tsushin Corporation
IEEE	Institute of Electrical and Electronics Engineers
IETF	Internet Engineering Task Force
IIA	independence from irrelevant alternatives
IP	Internet Protocol
IPR	intellectual property right(s)
ISDN	integrated services digital network
ITS	Intelligent Transport System(s)
ITU	International Telecommunication Union
ITU-T	International Telecommunication Union Telecommunication Standardization Sector
JAXA	Japan Aerospace Exploration Agency
JETRO	Japan External Trade Organization
JFTC	Japan Fair Trade Commission
JMA	Japan Meteorological Agency
JR	Japan Railroad East
JSMR	Japan Shared Mobile Radio
JUNET	Japan University Network
KDD	Kokusai Denshin Denwa Co. Ltd
LAN	local area network(s)
LDP	Liberal Democratic Party
LEO	low earth orbit
LRIC	long-run incremental cost
M&A	merger(s) and acquisitions(s)
MCA	multi-channel access
METI	Ministry of Economy, Trade and Industry
MIC	Ministry of Internal Affairs and Communications
MIDI	Musical Instrument Digital Interface
MIDP	Mobile Information Device Profile
MITI	Ministry of International Trade and Industry

MLIT	Ministry of Land, Infrastructure and Transport
MML	Mobile Mark-up Language
MOF	Ministry of Finance
MPHPT	Ministry of Public Management, Home Affairs, Posts and Telecommunications
MPT	Ministry of Posts and Telecommunications
MSAS	Multifunctional Transport Satellite-based Augmentation System
MTSAT	Multifunctional Transport Satellite
NB	narrowband
NCC	new common carrier
NICT	National Institute for Information and Communications Technology
NTT	Nippon [or Nihon] Telegraph and Telephone
NTTPC	Nippon Telegraph and Telephone Public Corporation
Ofcom	Office of Communications [United Kingdom]
PDA	personal digital assistant
PDC	personal Digital Cellular
PHS	Personal Handyphone System
PLC	Power Line Communication
POTS	plain old telephone services
QZSS	Quasi-Zenith Satellite System
RFID	radio frequency identification
RP	revealed preference
RPM	revealed preference method
RTK	real-time kinematics
SA	selective availability
SBB	Softbank Broad Band Corporation
SDP	Social Democratic Party
SMS	short message service
SP	stated preference
SPM	stated preference method
SSO	standard-setting organization
TACS	Total Access Communication System
TAI	International Atomic Time
TBL	Telecommunications Business Law
TCP/IP	Transmission Control Protocol/Internet Protocol
TDMA	time division multiple access
TELRIC	total element long-run incremental cost
3G	third generation
TTC	Telecommunications Technology Committee
2G	second generation
URL	Uniform Resource Locator
USTR	US Trade Representative
VICS	Vehicle Information and Communication System

VoIP	voice-over Internet Protocol
VRN	virtual reference station network
W-CDMA	wide-band code division multiple access
WIDE	Widely Integrated Distributed Environment
WRC	World Radio Communication Conference
W3C	World Wide Web Consortium
WTP	willingness to pay

Introduction

Ruth Taplin and Masako Wakui

The telecommunications market in Japan continues to be one of the largest-scale in the world and has reached a high technical standard. The market value of Japan's information and communications industry reached ¥116 trillion in 2002. At the end of fiscal 2003, the number of subscribers to telephone services was 51.59 million, the number of mobile phone subscribers was 81.52 million, and more than 77 million people were using the Internet. Broadband service is penetrating steadily and quickly. As of September 2004, there were about 12.8 million ADSL (asymmetric digital subscriber line) subscribers and two million FTTH (fibre-to-the-home) subscribers. CATV (cable television) has not been very successful in Japan, but even so has approximately 2.8 million subscribers. The coverage of optical fibre network installed reached 80 per cent by the end of fiscal year 2003.

In Japan, over a long period of time, substantial efforts and trial and error have made for an accumulated wealth of experience which allows the majority of people to enjoy a high standard of telecommunication services. The history of modern telecommunications in Japan started in 1854, when United States commodore Matthew Galbraith Perry brought telegraph machines to the people. Japan had been a closed country for over two hundred years at that time and rejected almost any trade and communication with foreigners. Under pressure from Western countries keen to broaden their markets and territories, Japan reluctantly opened its borders. At that time, technology in Japan was far behind that of Western countries, as it was in most Asian countries. Realizing that the control and leadership rested with advanced Western countries, the Japanese saw that modern industry and advanced technologies were critical for Japan's survival as an independent country. Japan started drastic reform under the government leadership, but in close cooperation with industries. Telegraph and telecommunication services were one of the technologies that began to spread during this period. The commercial telegraph service started in 1869, one year after this change, and telecommunications started in 1890, a year after the first modern Japanese written constitution was published.

Since then, Japan has experienced two world wars, wholescale economic destruction, radical democratization, and economic development, which was once called the 'Japanese miracle'. This spectacular economic growth did not last for ever, however. After the unprecedented prosperity was revealed to be a bubble without solid foundation, Japan had to go through an economic downturn during the entire 1990s, now referred to as the 'lost decade'. Japan is now struggling to regain its confidence through structural change of its political, legal and economic systems.

Japanese telecommunications is one of the sectors that have undergone drastic reformation in such structural change. The biggest change since the Second World War is, of course, liberalization and privatization. New laws passed in 1984 eliminated monopoly in the form of NTT (Nippon Telegraph and Telephone) and established a set of basic rules for the liberalized market. In 1986, three competitive carriers, or new common carriers (NCCs), started business with a network between the Tokyo–Osaka industrial corridor. Liberalization in 1984 was accompanied by privatization, and a new NTT was incorporated as a private company. Since then, typical competition-promotional measures and regulation have been established.

After these liberalizations in Japan, how has competition been promoted? What type of achievement has been attained? Although these liberalizing and pro-competitive reforms and measures have reached a certain level of achievement, the historical path has not been smooth. There are still many issues left unresolved, and this book charts the progress and current issues of the telecommunications market and policies in Japan.

Structural reform to date

In Japan, full-scale liberalization of telecommunication services started in 1985–1986 by abolishing the statutory monopoly of Nippon Telegraph and Telephone Public Corporation (NTT Public Corp.) and giving permission to two NCCs. Giving permission to providers of services was never going to be sufficient to bring competition into the telecommunications market, because NTT could use its dominant position to block the entrants. The incipient competition in telecommunication service markets needed regulations dealing with the former monopolists' exclusive actions. In particular, it is critical to ensure access to the 'last mile', or the telephone line from the premises of subscribers to the local exchange, which is the most difficult part for entrants to build up.

The Japanese telecommunications industry is under sector-specific regulation, and the Minister of Internal Affairs and Communications and the ministry (MIC) are in charge of this regulation. MIC is the successor to the Ministry of Posts and Telecommunications (MPT) which in English was called the Ministry of Public Management, Home Affairs, Posts and Telecommunications (MPHPT) until 2004. The Telecommunications

Business Law (TBL) regulates telecommunications carriers' business practices, including conduct affecting competition. Under the law, the Minister of Internal Affairs and Communications retains the broad power to issue an order to improve business (TBL Art. 29), which can be exercised to facilitate competition.[1] Antimonopoly law is also applicable to anticompetitive behaviour.

In addition to the above, the dominant company is particularly regulated by the 'Designate Facility System Regulatory System', which was introduced in 1997.[2] Under this system, the telecom operators who install the 'designated facility', which is the facility access that is critical to competitors to provide any meaningful service in the market, is regulated strictly. They are particularly obliged to obtain authorization of interconnection tariff *ex ante*, publish technical specifications, not to discriminate unfairly, and so forth. The facility of NTT West, NTT East and NTT DoCoMo has been designated as such.

The regulation tries to ensure access to the essential network facilities of dominant carriers. Under the TBL, any telecommunications carrier must agree to a request for interconnection to its telecommunications circuit facilities, except for a number of listed reasons, such as hindering the smooth provision of telecommunications services (TBL Art. 32). If no agreement is reached through private negotiation, operators can use a dispute resolution scheme supported by the government as the last resort (TBL Art. 35). As we have seen, the carrier installing the designated facility has to set the tariff and get the approval of the Minister *ex ante*. Under the law, the Minister must approve the interconnection agreement tariff when it satisfies the specified requirements, one of which is that the interconnection charge is based on the cost. The Minister may set the standard interconnection charge based on a 'forward-looking long-run incremental cost' (LRIC), compliance with which should mean satisfying conditions for the approval. Some interconnection charges are indeed based on LRIC.

At the same time, other commonly used competition promotion measures, such as a 'carrier pre-selection system', were introduced (in the case in 2001). Number portability among fixed networks has been gradually introduced. Number portability among mobile companies is planned for introduction in 2006, a move set to bring harsh competition in mobile markets.

In the process of liberalization, the dominant telecommunications companies have been reorganized so that their ability and incentive to hinder competitors' growth through using their dominant power over some markets are diminished. AT&T was split into several companies to provide local telephone services, and AT&T now provides the long-distance service. In the United Kingdom, BT has been operationally separated. In Japan, after heated discussions on the need to compete with the US-style divestiture, NTT was reorganized as a group of corporations controlled under NTT holding companies (in 1999). Now NTT group companies are organized under a NTT holding company. NTT East and West provide regional telecommunication services and own bottleneck local loops. NTT (the holding company), NTT East and West come under the 'Law Concerning Nippon Telegraph and Telephone Corporation, Etc.' (the so-called NTT Act), which specifies the business and obligations of the companies. NTT West and East are currently under business-line restriction, and would need to get special permission to do business in other intra-prefecture telecommunication services. So far they have been granted permission to provide inter-prefecture voice-over Internet Protocol (VoIP) services. Permission is normally accompanied by a series of conditions, such as fair treatment of competitors and firewalls among different service departments. The NTT group also has other prominent companies, such as NTT Communications (long-distance and international), NTT DoCoMo (mobile) and NTT Data (data communication).

Drastic change

The reform of these regulations fundamentally transformed the telecommunications market. In the telecommunications sector, in combination with this reform, drastic technological innovations were instituted, which also transformed the nature of the telecommunications network and services. Among other transformations was the spread of the Internet on a massive scale. In a network that is beholden to Internet protocol, the network and services are operated through the combined efforts of different elements controlled by the numerous entities, which makes a sharp contrast with conventional telecommunication systems that are integrated vertically. This transformation changes the nature of relationships between companies and the nature of competition among them. Another change is caused by the technological need to distribute ever larger contents, which has transformed the relationship between the broadcasting and telecommunication businesses.

In Chapter 1, these changes in regulations and technology in the telecommunications sector will be analysed by using the concepts of 'modularity' and 'interface'. The term 'module' refers to a component of goods and services that is designed independently, while 'interface' means an interface between these modules and shows how modules are connected technically or in the context of business. The structural changes in the telecommunication market can be explained in terms of changes between these modules and interfaces. For example, the regulation of interconnections is meant to design the 'interface' between network components as modules by rules. Content distribution services over the broadband can be understood as the Internet adopting the combination of open interface and content. In Chapter 1, structural changes and current conditions in the telecommunications sector since liberalization are clearly analysed using the concepts 'module' and 'interface'. Future prospects are also analysed concerning the telecommunications industry and competition policies.

Market conditions

Now, numerous companies are competitively offering services of a high standard. Table I.1 shows the major companies that are currently operating in the Japanese telecommunications market. It demonstrates that NTT, the former statutory monopolist, the KDDI Group and the Softbank Group are in a competitive relationship in many markets. KDDI is a company that was formed as a result of a merger between KDD, which used to monopolize the international telecommunication service market under the law, and IDO and DDI, which entered into the market after liberalization. Softbank is a company which has its origin in the computer and information technology market. It became a leading player in the telecommunications field that has secured a substantial market share through offering Internet provider services, especially ADSL, and the Internet telephone (VoIP) services over ADSL. Recently, it has expanded its business into the traditional telecommunications sector such as fixed-line and international services through the acquisition of Japan Telecom and Cable & Wireless. It is also planning to enter the mobile

Table I.1 Major players in the telecommunications service market

	VoIP	*Fixed*		*Mobile*
		Regional	*Long-distance and international*	
NTT Group	NTT West and East NTT Com (FLET'S and plala, OCN, etc.)	NTT West NT East	NTT Communications	NTT Docomo
KDDI Group	(DION)	KDDI		(au, Tu-Ka)
Softbank Group	SoftBankBB (Yahoo! BB)	Japan Telecom		
Others	Fusion [Powered Com Gr.] K-Opti.com [Kansai Electric Power] USEN	Fusion [Powered Com Gr.]		Vodafone Willcom (PHS)

Source: Author's analysis.

telecommunications market through the spectrum that is to be allocated to new entrants in the near future. In addition to these three leading groups, power companies own fixed networks and are providing telecommunication services. In the mobile telecommunications market, Vodafone, which is a multinational company with its headquarters in the United Kingdom, established its position parallel to the two major Japanese companies of NTT DoCoMo and KDDI (au). Cable TV companies also operate in the Internet provider service market, especially in the broadband sector.

How is competition conducted among these various players? What type of position, compared with NTT, do competitors and NCCs share in the market? What kinds of services are users enjoying? In Chapters 2 and 3, two highly developed and economically important markets are analysed, namely the broadband and mobile telecommunications markets.

Chapter 2 deals with the broadband market. The spread of the Internet in Japan was initially delayed. Indeed, the delay was once considered to be an obstacle to the economic growth of Japan. However, recently the spread of the Internet has progressed rapidly, largely owing to such factors as the expansion of the Softbank Group and measures implemented by the government. Currently, Japanese users can enjoy a high standard of service at very low cost, a phenomenon that is the subject of world attention.

The rapid spread of this high-standard service and the development of competition in the broadband market have contributed towards the creation of a new set of difficulties, however. It has become difficult to understand the market status clearly when so many services are being offered by business operators, but such an understanding is necessary in order to articulate and enforce regulations effectively. Under commission by MIC, a regulator, the author of Chapter 2, Takanori Ida, developed market review tools that can analyse such complicated statuses in the broadband market. In Chapter 2, by using the framework and tools obtained for his study, he explains the market status and conditions of competition. Issues for future policies will be also examined, using these insights.

Chapter 3 deals with the mobile telecommunications market. Those who travel through Japan often note people checking their road map, making reservations and enjoying high-quality music/movies and games on their mobile phones. In Japan, the photo/video function spread quickly and is now a standard feature of all mobile phones. Japan appears to be following a different path of development from that of other countries. In fact, many of these services and devices were originally uniquely developed by Japanese companies and, in many cases, their use and availability were limited to Japan. On the other hand, however, now Nokia phones are available, Vodafone has established a base in Japan, some mobile phones have the Qualcomm logo on the back, and the same standard is adopted by Europe and NTT for third-generation mobiles. It appears therefore that the Japanese market is becoming globalized. Chapter 3 analyses the globalization of the Japanese mobile market, its process and effect. The study also reveals the impact of technological changes.

Licensing and spectrum

As has been mentioned already, the number of business operators in the telecommunications market in Japan is constantly increasing and competition among them becoming fiercer. In the past, when only NTT and KDDI were allowed to operate telecommunications services, competition could not exist. This transformation has occurred since the government initially changed regulations concerning market entry.

The regulatory burden on companies wishing to enter the telecommunications market has been constantly relaxed. When liberalization began in 1985, the government retained the idea of 'excessive competition', or, in other words, a concern that excessive competition would damage the prospects of building and maintaining a high-quality telecommunications infrastructure. This reflects the provision in the TBL that sets the existence of sufficient demand and prospective inexistence of excessive facilities as obligatory licensing conditions by the Minister (TBL ex. Arts 10(1) and (2)). The provision was finally repealed in 1997.

In 2004, the regulatory burden on those entering the market was further reduced through replacement of the governmental permission system with a new registration and notification system. Previously those who provided a telecommunications service by establishing a public transmitting and switching facility were required to obtain permission by submitting a detailed business plan, prospective financial status and other details.[3] After the amendment, the permission system disappeared. The regulation scheme is generally moving from a formal categorical entry permission system to one of individual conduct rules.

There are still certain cases in which business operators are required to engage in prolonged and complicated negotiations with the government in entering the market, however. For example, operators are required to obtain a spectrum allocation to conduct business. Spectrum is a limited resource and many countries that suffer from such limitations are now having operators competing for its allocation. Its allocation will have a significant impact on the future course of the development of the mobile telecommunications market and the type of competition.

To effectuate more efficient use of spectrum, many countries have conducted various reforms, such as introduction of spectrum auctions and spectrum trading. In Japan, although certain reforms have occurred, the basic structure stays the same, which is the MIC command and control system.

Chapter 5 explains the spectrum allocation system and its history and background in Japan. The author, Hajime Oniki, shows lucidly how spectrum policies are determined and implemented in Japan and how policymakers and other players behave during the process. Readers will learn how spectrum should be allocated and under what circumstances more effective spectrum policies might possibly be realized. The chapter contains the most recent material and analyses for those who are interested in telecommunications policies and administrative dynamics.

Satellite

Even now that private companies offer telecommunications services, and competition has facilitated further development, there are still certain cases in which the government is required to play a leading role through its involvement in fund release or project planning and drafting. One such case where the government is expected to play a role is in satellite development. Satellites that are used for military and security purposes and that utilize cutting-edge advanced technology for the purpose of research and development are still at present developed and managed under government leadership, mainly by the national space agencies such as in Japan or by the military – in the US case by the Pentagon. This trend will change, however, with the civilian-operated Galileo GPS system of Europe.

Satellite systems are crucial to the development of modern telecommunications, and communications in general. Satellites that can offer the highest standard of Global Positioning Systems (GPS) will allow the widest and most important applications for society, ranging from improved emergency services to remote location and seabed detection systems. Improved telecommunications services mean that not only can impending natural disasters be spotted quickly, but so too can certain kinds of human rights abuses, which can then be beamed across the whole world, preventing genocide, for example, as the whole world will be aware of such attempts within minutes and governments working together can stop such unacceptable behaviour rapidly.

The countries of the world are becoming increasingly divided according to what satellite/GPS system they will support and use, as access to vital information is essential to a modern globalized system. This division brings satellite systems into the forefront of geopolitical relationships, which will have great implications for the future. Japan is developing its own terrestrial-based Quasi-Zenith Satellite System, which will continue to be linked to the American GPS system. This appears to be a geopolitical decision, as the Japanese attitude to Galileo GPS is unclear despite it, being a superior system to the American one. China, which has seen recent friction with Japan, has already made a financial investment in Galileo GPS.

In Chapter 4, we see how the importance of satellite systems to Japan is shown by the great efforts made to develop the Quasi-Zenith Satellite System. Its *raison d'être* and applications in relation to society and economy are explained. The geopolitical alignment of countries between the American GPS and the European Galileo GPS system will be discussed and analysed in relation to the pivotal role of Japan. These issues are analysed not only within the context of the economy and business but also from the geopolitical viewpoint: is Japan likely to join the Galileo Project? If not, what is the likely scenario? How significant is the involvement in this project (or non-involvement) to Japan? Will the recent close alliance between China and India use satellite technology to help change the global power structure?

R&D and intellectual property

Chapter 6 deals with issues concerning R&D and intellectual property in the telecommunications sector. The establishment of R&D in telecommunications has been developed uniquely in each country, by reflecting differences in relationships among companies and conditions of competition and history. NTT has had a decisive influence on telecommunications technologies and standards in Japan, as the largest facilities procurer and the dominant service operator with the largest number of users. The position has been supported by a group called a 'family' consisting of equipment manufacturers in close contact with NTT. NTT and these manufacturers have cooperated and conducted joint research and development projects. However, it seems that this type of R&D system is about to change as a result of the structural changes mentioned earlier. In Chapter 6, the following issues are discussed: how technology development has been conducted in the telecommunications sector in Japan, how intellectual property has been owned and utilized, and the types of changes that have been occurring. Measures and policies encountering these changes are also reviewed, which include patent pools, standard-setting bodies' intellectual property rights (IPR) policies and regulations under Antimonopoly Law.

Politics in telecommunication

The Japanese telecommunications industry is under sector-specific regulation, and MIC is in charge of this regulation. The Japanese telecommunications regulatory structure has been described as an 'interesting anomaly', given the state of its national network and levels of competition.[4] There is no independent regulatory body, such as the Federal Communications Commission (FCC) in the United States or Office of Communications (Ofcom) in the United Kingdom. In Japan, there is no structural separation between the industrial promoting function and regulation. Their mandate is broadly and vaguely specified as 'securing and advancement of proper and smooth electronics communication' and 'the matter concerning development, improvement and coordination of telecommunications business' (Arts 3 and 4 of the Law Establishing MIC). Their role indeed ranges over licensing, regulation, universal services, research policies, spectrum management and international cooperation. Enforcement of competition rules is a part of their business. The minister is given authority to permit the entry, authorize the spectrum use, regulate carriers' conduct and protect consumers. The minister is appointed by the Prime Minister, and is normally a member of the Diet. A colleague in the Diet, the Finance Minister, once owned about a 50 per cent share of NTT (about 34 per cent at the time of writing).

Under the system, the politicians were suspected of exercising their power to exclude foreign companies and protect NTT. In fact, the path that Japan has taken in reforming regulations has not been a smooth one, nor has it been particularly simple. In the telecommunications sector, a so-called iron triangle, meaning a stable and closed relationship among industries, ministers and politicians, which is seen as a feature of the Japanese political and economic system in other sectors, does not exist. Politics with regard to telecommunications in Japan has remained complicated, with political bargaining particularly between those who protect NTT and those who promote active competition, especially the NTT trade union; among NTT supporting groups, users of telecommunications services and companies from other sectors that depend on the development of the telecommunications services; between MIC as a regulator and the Ministry of Finance (MOF) as a shareholder of NTT; between the Japanese government and foreign governments, between the ruling party and opposition parties. Furthermore, none of these organizations is monolithic. The foreign governments do not necessarily reflect all the time only the interests of foreign companies, and were those that were actually supported by competitive operators in Japan.

Chapters 7 and 8 examine the dynamics in creating telecommunications policy. In Chapter 7, this kind of complicated political process is analysed by using the concept of policy network. The approach assumes that public policy is mainly the product of the interaction of several public and private stakeholders specific to its field, and that the mapping of those players in a network-like format provides a vehicle for understanding the structure of the policy-making process. By following the policy discussion on the organizational reform of NTT from 1981 to 2003, the present study examines how the structure of the policy-making process has changed for that period. The examination indicates that the policy output often depends on how relevant stakeholders make alliances and interact with each other, and that the structure of such networks is far from constant. One point that should be noted as a result of the analysis is that NTT has moved more towards the centre of the policy network. The study concludes with a discussion on institutional arrangements and the operation of competition policy within that context.

In Chapter 8, international talks between the Japanese and US governments with regard to telecommunications are analysed. The topic to be dealt with is access charges. NTT obliges competitive operators to pay when its regional communication network is mutually connected to the networks owned by such operators. For a long time, NTT's local access charges, or the cost to interconnect NTT's local exchange, were high, once estimated at three to four times higher than the prices in the United States and European Union. In the United States and European countries, the above-mentioned 'LRIC' access charge scheme was introduced by regulators to set the charge at a reasonable level. After a harsh controversy, the LRIC was finally introduced in Japan because of widespread criticism, intellectual support from academics, the sense of risk to the government that the charge was blocking the growth of affordable telecommunications services with economic growth as a whole relying on it, and also because of the pressure from foreign countries. Internationally, NTT's expensive access charge was severely criticized as a barrier to service trade.

In Chapter 8, the US–Japan negotiation over the access charges between 2002 and 2004 is analysed: how MIC, as a negotiator, behaved towards the US government, what determined MIC's behaviour and negotiation process, what the series of talks brought to markets and regulations in Japan. The author, Motohiro Tsuchiya, considers that players in international politics are not those who can insist on their own interests independently and consistently, but rather those who behave by adjusting and guessing the interests of each group within the country and exercising their own influential power against such groups according to the domestic political circumstances at the time. In addition, misunderstanding about their own interests, the interests of counterparts and their own power to influence can be a factor leading to negotiations to fail. From this point of view, the relationships between domestic and international politics over access charges are analysed in detail. Insight into issues on future international negotiations are given as well.

Continuous challenge

How will the telecommunications market and regulations in Japan change in the future? What will be the main issues? In the telecommunications sector, it is likely that technological innovation will continue and competition and business will go through a further transformation. Some topics such as the spread of optic fibre networks (Chapter 2), the development of mobile telecommunications (Chapter 3) and the contents distribution business that are utilized within these infrastructures (see Chapter 1) are analysed. These technological innovations and changes in market status will continuously lead the transformation of the industry. Another factor changing the market is globalization, as we see in Chapter 3, which analyses the impact that international standardization and foreign companies have on mobile telecommunications in Japan, and in Chapter 4, which details the current conditions and future prospects of the satellite system that have an impact on a global scale. Globalization also affects technological development and how the owning and utilization of intellectual property will develop, as we see in Chapter 6. Telecommunications policies and regulations in Japan will be continuously required in addition to appropriate measures being introduced and, if necessary, existing systems and regulations being restructured to respond to changes in these technologies and markets. Spectrum is one of the issues that require an urgent response (see Chapter 5)

It is reasonable that interests and players interact in a complicated manner and government organs have more difficulty in exercising decisive influential power in terms of telecommunications policies and regulations. (see Chapter 7). They are facing difficulties in international negotiations because it is impossible to make a precise judgement on the national interest (Chapter 8).

It is therefore necessary to review and examine what type of system is the most effective to realize the appropriate policies and regulations. In Japan, there has been criticism that telecommunications are governed and regulated under an obscure system lacking transparency as well as a capable and independent regulator. From this point of view, it is worth noting what role the Fair Trade Commission can play in the future (see Chapter 7). Similarly, the role of the courts is under focus. The courts have never taken a significant role in telecommunication regulation in Japan. Private action under the Antimonopoly Law (Competition Law) and against the government has been always possible. Yet it is rarely used, as if it did not exist at all. The relative lack of court action does not apply only in the telecommunications industry. Basically, the Japanese do not tend to go to court, mainly because legal action is quite time-consuming and requires a lot of money, there are few lawyers who are familiar with sector-specific regulation, and the results of disputes are unpredictable as there have been no legal precedents. Expertise and information have tended to be monopolized by the big leading companies and the government, as noted in Chapter 1. Despite these difficulties, however, five competitive telecom operators, including KDDI, Japan Telecom and Cable & Wireless, sued the MIC in 2003, as noted in Chapter 8. This ground-breaking news was followed by a shocking lawsuit by Softbank against MIC alleging that spectrum was unfairly allocated so that there was no space for new entrants to start mobile businesses.

MIC itself has introduced various measures to improve transparency. The public consultation system was implemented, the amount of published data and information has been significantly increased to the extent that MIC said that it has become the most transparent department among Japanese ministries, and the policy and administrative evaluation system was introduced. Expertise from outside the Ministry is increasingly used. One example of its use is the council system and study groups, prominent members of which are academics. Another example is the market review, which was conducted in cooperation with scholars and universities, the result of which are explored in Chapter 2.

Although it cannot be said these reforms are sufficient, there is one where progress has certainly been made. It is substantially easier than before to access information on policies and markets. The amount of information disclosed by MIC has increased, but also the quality of information has been enhanced as researchers become involved in information production processes, as seen in the market review (Chapter 2). A further result is that the improvement in quality and quantity of the information has activated discussions on policies by third parties such as researchers. Although the quantity is still small, information disclosed in English has also increased, which will help foreign entities examine Japanese markets and policies from a new perspective and in a critical manner, by taking into account their experiences in other countries. The litigation is making public facts which are not normally disclosed. To bring regulations and policies to light and expose them to continual examination is the best measure to ensure that policies and regulations are impartial and progressive, which will be always the case regardless of what directions the technologies and markets take.

Notes

1 The order can be issued particularly when 'there is a risk of extreme impairment of the public interest because proper operations of other telecommunications carriers are interfered with, due to the fact that the telecommunications carrier unfairly discriminates against a specified telecommunications carrier in interconnecting or sharing telecommunications facilities or in providing wholesale telecommunications services, or to the fact that the telecommunications carrier conducts other unfair operations concerning such services' (TBL Art. 29(xi)).

2 Fixed telecommunications facilities and mobile facilities fall under a different scheme. In the fixed telecommunications market, facilities deemed as 'essential for the enhancement of the users' benefit and the comprehensive and rational development of telecommunications' are designated as such (TBL Art. 33), The regulation includes interconnection tariff *ex ante* authorization, which covers access charge regulation and publication of technical specifications and, concerning the fixed telecommunication service facilities, account separation (TBL Art. 33). They are also obliged not to discriminate unfairly, not abuse proprietary information regarding other telecommunications carriers or users, not unduly compel or intervene upon the business of other telecommunications companies, and to have separated accounts (TBL Art. 30). NTT West and NTT East come under the regulation. In a mobile market, the facilities that ensure 'appropriate and smooth interconnections with the telecommunications facilities of other telecommunications carriers' are designated, and those who install the facility have to notify the detailed interconnection tariff in advance (TBL Art. 34). When the market share of the company with the facility exceeds a certain

amount, it is placed under additional regulation that prohibits unfair discrimination and so forth (TBL Art. 30). NTT DoCoMo comes under this regulation.

3 After entry, they became a 'Type 1 Business Operator', as opposed to a 'Type 2 Business Operator', an operator without switching facilities, and the business remained under strict regulatory control. Legislation forbade them to cease production under their own authority. After the amendment, the distinction between Types 1 and 2 disappeared as well.

4 Lisa Suits, 'Lesson from Asia and the Pacific' in Ian Walden and John Angel (eds), *Telecommunications Law, 1st edn*. London: Blackstone Press, p. 457.

1 Changes in the interface and industry structure

Sumiko Asai

Introduction

Over the last two decades, the Japanese telecommunications industry has changed dramatically from a monopoly situation to a competitive environment. In addition to fierce competition among traditional telephone carriers, the Internet is today recognized as a new competitor in the telecommunications market. Regarding transmission facilities, narrowband public networks are in the process of being transformed into broadband networks using the Internet Protocol (IP). Thus, the Japanese telecommunications market is at present in a transitional phase.

This chapter surveys the changes in the market from the following three perspectives. First, the chapter analyses the industry structure of telecommunications in Japan, using the concept of modularity and the interface between modules. Up until now, discussion of Japanese telecommunications policy has tended to be confined within a small group of researchers. One reason for this is the existence of technical terms that are necessary for an understanding of telecommunications. Specialized knowledge of telecommunications is a barrier restricting the entry of researchers from the fields of law and economics. Another reason is the lack of information on policy decisions. Japanese telecommunications policy has generally been discussed in the councils and study groups established within government agencies. Until the early 1990s, the discussion in these councils made up of experts was not open, and information, except for the final reports and recommendations, was not available to non-council members. However, at present, the process of policy decision-making, including the discussion in the councils, has become open, and a lot of material is available over the Internet. The problem of information disclosure has been almost solved as of today, but a minimum level of specialized knowledge remains a requirement to examine telecommunications policy. In addition to this, recent technical innovation has accelerated change in the telecommunications markets, and this makes it difficult for an individual researcher to have a comprehensive understanding of the current situation. Therefore, I avoid the use of industry-specific terms in this chapter and consider the trend in changes in industry structure, utilizing the concepts of modularity and interface employed in the field of business administration.

Second, although several services such as third-generation cellular phones have more than one standard, connection of networks across countries is now a routine occurrence. Since many standards for public telecommunications networks have been established by the International Telecommunication Union (ITU) and other public institutes, the standards and technologies in the telecommunications market are common worldwide and represent open information that a third party can access. However, differences in the policy and regulations may exist between countries. When policymakers establish a new regulatory framework, the continuity of regulatory systems is generally taken into consideration. Even when large-scale regulatory reform is conducted, the reform is based on a review of the current system – that is to say, the reformation of a system is still influenced by the former systems. In addition, if a certain social arrangement were to be designed and operated, the relevant systems would be designed on the assumption that the arrangement would be maintained in the future. Therefore, once a system that is different from systems in other countries has been designed, the differences between the systems are not easy to resolve, even when technical standards have been established. Since path dependence is considered to be one factor in institutional diversity,[1] this chapter attaches importance to a historical description of Japanese telecommunications policy, which is the second perspective.

Third, this chapter deals not only with public networks, but also with the Internet and the content provided by telecommunications carriers. Before the development of the Internet and cellular phones, a number of studies examined plain old telephone services (POTS). However, POTS has now been partly replaced by cellular phone and IP services. Therefore, making a survey of POTS is not sufficient to ascertain the current status of the telecommunications market. On the other hand, telecommunications services are defined as the transmission of information to a specific person employing electrical facilities, and telecommunications carriers are not engaged in the production of information. However, several cases may be observed where the penetration of certain telecommunications services depends on whether the carriers are able to provide attractive content. For example, one factor in the penetration of i-mode, one of the cellular phone services provided by NTT DoCoMo, is the provision of a great deal of attractive content, and there is a positive feedback mechanism between cellular phone services and content.[2] We may say that the Internet and content have had an impact on the development of services in the telecommunications market.

When we look at the overseas markets, we can see a wide variety of regulatory frameworks in the telecommunications market. Such institutional diversity occurs as a consequence of changes in other industries such as the computer and content industries, as well as of technological progress, path dependence and institutional complementarity. Therefore, other fields adjacent to the telecommunications industry are also included as objects of consideration in this chapter. I shall now proceed to explain the concept of modularity and interface, which are key words in this chapter, then proceed to a discussion of Japanese telecommunications policy.

The concepts of modularity and interface

The production process for goods and services is generally divided into several modules and the interface between these modules. According to Baldwin and Clark (1997),[3] a module is defined as a component of goods and services that is designed independently, and dividing the production process along modular lines is defined as modularity. The term 'unbundling' in the telecommunications market has been used as a synonym for modularity. To take an example of telephone services, they are divided into several modules, including customer premises equipment (CPE), telephone lines and switching facilities. These modules are connected according to the technical standards established by the ITU and the Ministry of Internal Affairs and Communications (MIC).

Modularity has been discussed in the field of business administration, since Baldwin and Clark (1997, 2000) proposed the concept of modularity and analysed the computer industry in the United States.[4] In Japan, Fujimoto *et al.* (2001) analysed the automobile and other industries,[5] and Asai (2004) analysed the information industry using the concept of modularity.[6]

There are several ways in which modules and the interfaces between modules can be combined. First, a product may be composed of one module. This is an integrated system, and the production is completed within a single firm. Several studies have pointed out that IBM mainframes before the early 1960s are an example of the first case.

Second, the production process may be divided into more than one module, and the interface between these modules is specified within a firm or among the firms concerned. The interface is closed or proprietary information. Baldwin and Clark (2000) examined the production process of IBM System/360 as an example of the second case. IBM System 360 was a mainframe developed by IBM in 1964. The functions were allocated to physical components, but the interface between the components was designed by highly skilled staff members inside IBM.

Third, the production process may consist of more than one module and an open interface that a third party can access. Examples of the third case are a personal computer developed by IBM in 1981 (the IBM PC) and the Internet. Thus, IBM has been involved in all three types, reflecting its technological progress in the field. Japanese telecommunications networks have also been transformed from integration with closed interface into modularity with open interface, as mentioned in the following sections.

In the third case, firms are given the choice to 'make or buy'. That is, they may produce some modules themselves and procure other modules from the market in order to provide the final goods or services. When an interface is fixed with open information, firms do not need to take other modules and the interface into consideration. Therefore, the transaction cost incurred for combining several modules is reduced, and firms are able to concentrate their research and development (R&D) in certain fields. In addition, since firms do not need to produce all modules themselves, small-scale firms that do not have substantial assets may enter the business.

On the other hand, in addition to the low entry barrier, since modularized products are homogeneous, there is greater price competition. As a result, a decrease in the fees charged for telecommunications services improves consumer welfare. However, it also means that telecommunications carriers cannot maintain their profits unless they have other competitive advantages.

An open interface includes *de jure* and *de facto* standards. In the case of telecommunications, the interconnection rules that consist of economic agreements and technical standards have been prescribed by the government, and these rules are the *de jure* standards. On the other hand, the Internet is a worldwide network that uses the *de facto* standard TCP/IP (Transmission Control Protocol/Internet Protocol). Table 1.1 shows the combination of the production process – that is, integration or modularity – and open or closed interfaces in the information industry.

The combination of the production process and open or closed interface has an impact not only on the organization and management of firms, but also on regulations and competition policy. NTT Public Corp. and

Table 1.1 Selection of production process

	Closed interface	*Open interface*	
Integration	NTT Public Corp's Network Mainframe (before IBM System/360)		
Modularity	IBM's System/360	*De jure*	*De facto*
		NTT's Network CS Broadcasting	IBM-PC The Internet i-mode

Source: Sumiko Asai, *Zyoho Sangyo no Togo to Mozyuruka* (Integration and Modularity in the Information Industry) (Tokyo: Nippon Hyoron, 2004, table 1).

IBM mainframe computers were integrated with a closed interface (Table 1.1). In the case of integration with a closed interface, firms tend to have market power. In fact, NTT Public Corp. was a monopolistic carrier and IBM was a dominant computer producer. Therefore, NTT Public Corp. was regulated by the Ministry in order to avoid abuse of its dominant power. IBM was charged by the Department of Justice with attempting to monopolize the computer market employing the integrated system in 1969. The closed interface that was common to NTT Public Corp. and IBM's mainframe computers led to a lack of transparency in setting prices and abuse of dominant power.

On the other hand, since open interface ensures transparency in business activities and promotes entry and price competition, the possibility of anticompetitive actions has been reduced. Accordingly, we may say that moves towards an open interface lead to deregulation. We will now proceed to a discussion of the changes in the telecommunications industry, employing the concepts of modularity and interface.

Before the introduction of competition: the age of integration with a closed interface

Interconnection has been a major issue that Japanese telecommunications policy has attempted to address. However, policymakers were not concerned about unbundling and open networks before the introduction of competition. To illustrate, the following is a brief overview of the history of Japanese telecommunications.

From the inception of the telecommunications business in 1890 until the end of the Second World War, telecommunications in Japan were operated and controlled by the state. Before the business activities of telecommunications began, the government had discussed the possibility of whether private companies should provide telecommunications services. However, because the main users of telephone services were government agencies, and a large amount of investment was required to operate the telephone industry, the system of provision by the government was adopted in 1889.

Although telecommunications networks have expanded smoothly since 1890, Japanese networks constructed by the state were almost entirely destroyed during the Second World War. After the war, various kinds of social systems, including the regulatory framework for telecommunications, were reviewed, with the result that the Nippon Telegraph and Telephone Public Corporation (NTT Public Corp.) was established in 1953. Although NTT Public Corp. was a government agency, it was independent of the government in relation to personnel recruitment and funding.

There were several reasons why a monopoly system managed by a public corporation was adopted. First, prompt reconstruction of telecommunications networks was needed in order for the Japanese economy to recover. While a large amount of investment was required to provide telecommunications services, no private company with sufficient funds to construct networks nationwide existed at the time. In addition, it was considered that unified management under a national plan was the best way of reconstructing the networks quickly. Therefore, monopolistic provision by the public sector was maintained.

Second, POTS had been regarded as a universal service, and its provision to all customers at an affordable rate had been considered to be a basic requirement. If private companies were engaged in the telecommunications business, policymakers presumed that they would not provide their services in rural areas, with resulting regional gaps in telecommunications. Therefore, provision by the public sector was considered to be more appropriate to fulfil the obligation of providing a universal service.

Third, the telecommunications market was seen as being a natural monopoly due to the subadditivity of the cost function, although no econometric studies were conducted at the time. Therefore, it was considered that a monopoly would lead to greater efficiencies than competition would.

Overseas, the telephone markets in European countries in the 1950s were public-sector monopolies. Although the US telecommunications business was operated by private companies at the outset, a large private company, Bell System, actually enjoyed a virtual monopoly in the US telecommunications market. While the establishment of the regulatory framework of Japanese telecommunications was being discussed, these circumstances in foreign countries were also considered. As a result, the government decided to maintain the monopolistic provision by the public sector after the Second World War, and NTT Public Corp. was established. However, since a public provider with a monopoly did not have any incentive to reduce costs, a self-supporting system was introduced at NTT Public Corp.

NTT Public Corp. provided not only telecommunications services, but also jointly designed telecommunications facilities such as switching systems, networks and terminals, in conjunction with certain manufacturers. These manufacturers were NEC, Fujitsu, Hitachi and Oki Electric Company, and these manufacturers and NTT Public Corp. were referred to as the NTT 'Family', owing to their close ties. Technical characteristics of the networks were designed by companies belonging to the NTT family, and the interface was closed information. NTT Public Corp. procured a large amount of communications facilities from these four manufacturers,[7] reflecting the robust demand for telephone services, and these companies gained significant revenues stably. Although these manufacturers were independent of NTT Public Corp. from the standpoint of their organizations, NTT Public Corp. and the specific manufacturers were vertically integrated with a closed interface, to all intents and purposes. On the other hand, telephone users were obliged to lease their telephone terminals from NTT Public Corp. from the viewpoint of network reliability, and they could not select their terminals at their own discretion. To put it briefly, NTT Public Corp. provided users with integrated services, including the terminals.

Although the close relationship between NTT Public Corp. and these manufacturers was thought to contribute to R&D in Japan, the products designed by the NTT family were not compatible with foreign products. Since the NTT family had not been exposed to global competition in the communications facilities market and could not enjoy scale economies, Japanese products were more expensive than foreign products. In addition, incompatibility created a trade imbalance with regard to communications facilities, and foreign governments criticized the fact that NTT Public Corp. made exclusive deals with specific manufacturers. The procurement practices of NTT Public Corp. became the subject of discussion in bilateral and multilateral negotiations such as GATT (the General Agreement on Tariffs and Trade). As a result of these negotiations, in the early 1980s the networks were gradually opened. We may say that the networks of NTT Public Corp. were partly opened as a result of foreign pressure. However, NTT Public Corp. has been required to make further changes since the introduction of competition, as described below.

Introduction of competition without a change of interface

In April 1985, the introduction of a competitive mechanism as well as the privatization of NTT Public Corp. was implemented for the following reasons. First, the enormous subscription demand for telephones began to subside in the late 1970s. This meant that NTT Public Corp. had by this time almost completed the reconstruction of the Japanese telephone networks. On the other hand, NTT Public Corp., which had solved the pressing issue of network reconstruction, was now confronted with the problems of inefficient operations and illegal accounting manipulations during the late 1970s and early 1980s. It was pointed out at the time that the scale of NTT Public Corp. had exceeded the optimum size and its excessive size was one factor contributing to inefficiency and misconduct. Therefore, a competitive mechanism was introduced and NTT Public Corp. was privatized in order to improve its efficiency. However, in April 1985, NTT Public Corp. became NTT Corporation (NTT) without a change in the scale of the company, since no consensus on its break-up had been reached. Second, although the telecommunications market had formerly been considered to be a natural monopoly, technological progress and increasing demand enabled plural carriers to operate in the telecommunications market. Third, in the early 1980s the privatization not only of telecommunications carriers, but also of other public services providers such as the railways, was under discussion in the developed countries as a way to downsize government. In Japan, it was felt that deregulation and privatization of public utilities were needed in order to stimulate the economy and improve the efficiency of government activities. Fourth, it was considered that the monopoly of NTT Public Corp. could not respond effectively to diversified user needs, and the entry of new players into the market was required in order to provide a range of services at low rates.

When new common carriers (NCCs) entered the telecommunications market, they were confronted with several serious problems. The most severe of these was interconnection. While subscriber lines are essential to provide telecommunications services, constructing them is difficult and costly for NCCs. Therefore, subscriber lines constructed by NTT are called essential facilities, and NCCs wanted to interconnect with NTT's lines. However, NTT had no incentive to allow connection with other carriers since this would have led to NTT's losing its customers to its competitors. In addition, as many interfaces of NTT's networks were not open either technically or economically, and NTT's information disclosure was not sufficient, negotiations between NTT and NCCs on interconnection did not progress on a voluntary basis. Therefore, the Ministry of Posts and Telecommunications (MPT, now MIC) established interconnection rules and required that NTT disclose technical information and cost data in order to create a level playing field between NTT and NCCs, as described below.

Interconnection and the reorganization of NTT: from integration with closed interface to modularity with open interface

In 1987, three NCCs started their long-distance communications services by connecting with subscriber lines owned by NTT. However, since NTT was vertically integrated and information on segmented costs was not available, MPT was not able to set interconnection charges when NCCs began to provide their services. As a result, whenever long-distance carriers interconnected with NTT, they were obliged to pay NTT the same charges as users paid NTT for local telephone services until the interconnection charges were set in 1993.[8] In other words, NCCs were initially unable to set end-to-end service rates.

Japanese policymakers discussed the divestiture of NTT as well as its privatization in the early 1980s. However, consensus on its organization was not reached at that time, and the reorganization of NTT was still an open issue after privatization had taken place. Although MPT formally proposed the divestiture of NTT in 1990 to create a level playing field according to the NTT Law, NTT bitterly opposed the divestiture and again no agreement was reached. However, instead of the divestiture, the accounting separation of long-distance and local communications sectors was introduced into NTT. The first interconnection charges were set in 1993, utilizing the data derived from the accounting separation introduced in April 1992.

In the early 1990s, while the competition mechanism worked in relation to long-distance communications, international communications and cellular phone services markets, the local communications market was still presumed to be an NTT monopoly. From the late 1990s, several NCCs entered the local telecommunications market in addition to the long-distance communications market and requested NTT to connect their networks with NTT's local loops. In response to the request, MPT introduced unbundling of NTT's local loops by revising the Telecommunications Business Law (TBL) in 1997. According to the revised rule, NTT's networks were divided into several modules and its interface opened up. In other words, an integrated network with a closed interface was converted by regulations into a combination of several modules with an open interface. Unbundling gave NCCs the choice of whether they construct their network facilities or lease the facilities from NTT. The revised interconnection rules prescribed the introduction of interconnection accounts and a consultation process on interconnection as well as unbundling, and have been seen as having contributed to the development of competition and information disclosure.

On the other hand, the interconnection charges set by MPT since 1993 had been based on fully allocated accounting costs data submitted from NTT. However, the following problems regarding the calculation method used for interconnection charges still needed to be solved. First, since cost data are proprietary information of NTT, it is difficult for a third party to show that the submitted cost is accurate. In this respect, information disclosure was not sufficient to set the interconnection charge. Second, the cost may be overestimated owing to NTT's inherent inefficiency, as NTT was a monopolistic firm in the local communications market and did not have any incentive to reduce costs. In this case, the inefficiencies inherent in NTT are transferred to the other carriers that have connected with NTT through the interconnection charges. Considering these problems, in order to set the proper interconnecting charge, the MPT introduced an economic engineering cost model that did not depend on the accounting information provided by NTT in 2000. Although the model's reliability has been questioned, the engineering model has improved transparency in the decision-making process for interconnection charges as well as having reduced interconnection charges. The present interconnection charges have been set using the engineering model.

There are two approaches to changing an integrated network with a closed interface into an open network. One approach is the enforcement of interconnection rules, and the other is the separation of networks through the divestiture of carriers. However, the incumbent cannot be physically divided into a number of parts in the same way as networks are unbundled, because of the interconnection rules. Therefore, even when the incumbent is separated vertically, several interconnection rules are needed in order to create a level playing field.

Since the privatization of NTT Public Corp., MPT has asserted that the separation of long-distance and local communications sectors in NTT was essential in order to promote competition. Although NTT agreed to the separation of mobile communications and data processing sections from NTT,[9] it opposed the vertical separation of wired networks for telephone services. This situation was similar to the break-up of AT&T in 1984. However, at the end of 1996 MPT and NTT finally reached an agreement on the reorganization of NTT – that is, its separation into one long-distance company and two local-communications companies under a new holding company. The NTT long-distance company has been permitted to provide local communications and international communications services as well as long-distance communications services, since it has not held undue market power. Japanese telecommunications policy led not only to the unbundling of networks, but also to the reorganization of NTT. However, it is debatable whether the reorganization of NTT in 1999 was an example of separation or integration under the holding company.

Modularity with an open interface has progressed in the telecommunications market through the interconnection rules. The disclosure of NTT's information has thereby developed. In addition to this, small-scale carriers without many assets have entered the market. As a result, the number of Japanese telecommunications carriers exceeded 12,500 at the end of March 2004,[10] and users are able to select carriers themselves. However, the traditional telephone market has been dominated by a few large-scale carriers, and two-thirds of the total carriers in Japan are small-scale Internet access providers that play no leadership role.

Impact of the Internet as an open network on telephone services

The source of NTT's revenues in the 1980s was mainly POTS. However, the penetration of the Internet since the late 1990s has been significant, and some telephone services have been replaced by services using the Internet Protocol (IP). The Internet consists of several modules such as personal computers, access and backbone networks and servers. Since the communication protocol of the Internet is the *de facto* standard TCP/IP, the Internet has been divided into several modules with an open interface from its inception. This is in contrast to telephone networks, which at one time were integrated with closed information.

Large-scale investment is not required in order to provide Internet services, unlike telephone services, which need expensive switching facilities. As a result, numerous providers have entered the Internet services market. In addition, once users subscribe to an Internet access service provider and acquire email addresses from their provider, they are reluctant to change provider, because they do not like changing their addresses.[11] In other words, users are locked into the Internet service providers that they subscribed to first, and the Internet providers carry out aggressive marketing and the enclosure of users to acquire a competitive advantage, as can be seen from the case of Softbank Corp., which provides both ADSL and IP telephone services.

The tariff for IP telephone services is not linked to the distance of the calls, whereas the tariff for POTS changes in proportion to the distance involved. Therefore, IP telephone services are cheaper than POTS for long-distance calls. In addition, now that there is no obvious difference in quality between ordinary telephone and IP telephone services, users view IP telephone services as substitutes for POTS provided by common carriers, and some users have switched their telecommunications carriers from NTT to other carriers that provide IP telephone services. Therefore, telephone carriers are forced to provide IP telephone services even if they know that they stand to lose some of the revenues gained from POTS.

On the other hand, common carriers are able to reduce the amount they invest by replacing public switching networks with IP networks. At present, common carriers are in the process of converting into IP networks in response to both demand- and supply-side factors.[12]

Telephone services provided by public common carriers have been central to the telecommunications industry up until this point. However, computer technologies, including the Internet, have recently changed the telecommunications services and business strategies of common carriers. An examination of the computer industry and the convergence between computers and telecommunications is essential if the current status of the telecommunications industry is to be fully understood.

Convergence in the Japanese information industry; from the standpoint of module and interface

In Japan, when convergence of services in the information industry was discussed in the late 1980s and early 1990s, the main aspect of convergence was the relationship between telecommunications and broadcasting. Thereafter, the subject of convergence has expanded in proportion to the development of digitization and the Internet. Therefore, this chapter discusses convergence in the information industry including the Internet and computers.

Convergence has two aspects. The first is the combination of a module in one sector with a module in another sector. The second aspect is the emergence of intermediate services. While the first case occurs under modularity with an open interface, the appearance of intermediate services depends on the definitions of the relevant services; that is to say, the second aspect is related to the regulatory framework. First, the convergence of telecommunications and broadcasting discussed previously is examined in more detail. Then, content distribution over the Internet and convergence between telecommunications and computers are examined.

Convergence of telecommunications and broadcasting

In Japan, the *Dial Q²* service and direct broadcasting via a communications satellite (CS) have been cited as examples of convergence between telecommunications and broadcasting.

The *Dial Q²* service was introduced by NTT in 1989. This service provides information through telephone lines, and corresponds to the 900-number service in the United States. It consists of three modules: telephone lines as the transmission facilities, the collection of charges and the content provided by information providers. In this respect, the *Dial Q²* service, which combines telecommunications facilities and content produced by a third party, is an example of the first aspect of convergence. Since the NTT East and NTT West companies[13] collect information charges as well as dialling charges from their *Dial Q²* users, information providers do not need to collect charges independently, and are able to concentrate on producing content. This is one reason why many information providers have entered the content business. On the other hand, users pay NTT for both dialling charges and information charges, and do not have to make contracts with each information provider. Utilization without the need for a prior contract has promoted the penetration of the *Dial Q²* service.

In addition to information distribution, the *Dial Q²* service includes a so-called chat service that allows subscribers the opportunity to converse with non-specific users through NTT telephone lines.[14] While the TBL prescribes that telecommunications services are communications for specific persons employing telecommunications facilities, broadcasting services are defined as transmission of information intended to be received directly by the public, according to the Broadcasting Law. One of the *Dial Q²* services, the so-called chat service, is seen as an intermediate service in that it is a service for non-specific persons employing telecommunications facilities. Thus, the *Dial Q²* service also has the second aspect of conversion.

Although the demarcation between specific and non-specific users is ambiguous, the respective regulations for telecommunications and broadcasting are different.[15] Therefore, policymakers are obliged to classify intermediate services into either telecommunications or broadcasting. In 1998 and 2001, MPT published guidelines dealing with the convergence of telecommunications and broadcasting and presented concrete instances of service classification in the guidelines.[16]

Communication satellite (CS) broadcasting services can be used as an example of convergence between telecommunications and broadcasting. Since 1984, NHK has provided broadcasting services via a broadcasting satellite (BS) as well as terrestrial broadcasting, and had 14 million BS broadcasting subscribers as of March 2004.[17] While the utilization of BS is limited to the transmission of broadcast programmes, CS may be used for both telecommunications and broadcasting. In the light of the potential for

CS utilization, MPT revised the Broadcasting Law in order to begin broadcasting services via a CS in 1989. As a result, CS broadcasting services have been provided since 1992, and the number of CS broadcasting subscribers in Japan was 3.64 million at the end of March 2004. CS broadcasting is composed of the communications satellites launched by telecommunications carriers and programme content produced by a third party. CS operators for broadcasting are defined as facility-supplying broadcasters and programme producers for CS broadcasting are defined as programme-supplying broadcasters, according to the Broadcasting Law. That is to say, facility-supplying broadcasters are broadcasters as well as telecommunications carriers.

CS broadcasting is vertically separated into hardware (transmission facilities) and content. Since programme production generally does not require a large amount of assets, more than 100 programme-supplying broadcasters have entered the satellite programme market in Japan. It can be shown that the vertical separation into hardware and content has contributed to an increase in content, which is essential to sustain multiple channels.

Finally, let us take another example of convergence between telecommunications and broadcasting. Considering the development of broadband networks, the Law on Broadcasting Employing Telecommunications Facilities came into effect in January 2002, and operators who do not own network facilities have been authorized to distribute their content as broadcasters. We may say that the introduction of this law is a response to the convergence of services and the development of broadband networks. Indeed, several content holders have distributed their content by leasing ADSL and optical fibre networks from telecommunications carriers. On the other hand, ADSL providers that transmit television programmes expect to increase their number of subscribers and enhance their brands, and this regulatory framework has conferred benefits on both content holders and telecommunications carriers.

Recently, various types of content such as feature films have been distributed over the Internet. In this respect, the range of distributed content has expanded. Next, the distribution of content through networks is considered.

Content and its distribution

Dial Q^2 and CS broadcasting are examples of the separation of hardware and content, and this structure has been applied to content distribution over the Internet. One example of the content distribution service over the Internet is i-mode service provided by NTT DoCoMo. NTT DoCoMo introduced i-mode – that is, a packet data communications service using a cellular phone – in 1999. The functions of i-mode service are divided into three modules: voice transmission, the Internet access service and content on websites. For content, NTT DoCoMo has relied on content providers and adopted the *de facto* standard HTML (Hypertext Mark-up Language) for producing content. The adoption of such an open interface enables a lot of content providers to enter the content market. In addition, NTT DoCoMo has established criteria on the content available from an i-mode terminal, and has selected official sites, according to the criteria. i-mode users have easy access to official sites, and NTT DoCoMo collects information charges on behalf of the content providers of these official sites. In this respect, content producers authorized as official sites have a competitive advantage as compared with producers of non-official sites. Therefore, content providers endeavour to obtain official site status, and the competitive mechanism among content providers has worked. As the number of attractive websites increases, the utility of i-mode users also increases, as mentioned before. Thus, a positive feedback mechanism has worked between websites and users, and the combination of modularity and an open interface has resulted in the success of i-mode.

On the other hand, digitization makes it possible to store and reproduce content without deterioration in quality. Some publishers have not only published books, but also distributed their content over the Internet in the form of electronic publishing in Japan. In the case of narrowband networks, distributed content is limited. Recently, in proportion to the development of broadband networks, it has become possible to distribute not only sound and characters but also pictures over the networks. For example, feature films are distributed over the Internet, and we will be able to enjoy digital television programmes over the Internet or a third-generation cellular phone. As the broadband networks have gained greater market penetration, the range of content distributed over the networks has expanded.

In the case of content distribution services over the Internet, several Internet service providers (ISPs) collect information charges as well as their Internet access charges, similar to the *Dial* Q^2 service. In addition to the collection of information charge, for the Internet, both communication protocol and language for producing content are open interface, and producers can distribute content over the Internet without the need to make any form of prior arrangement. Therefore, we may say that modularity and an open interface have promoted content distribution over the Internet.

Convergence between telecommunications services and computers

Intermediate services appear not only between telecommunications and broadcasting, but also between telecommunications and computers. In the United States, intermediate services between telecommunications and computers have been the subject of discussion for a long time. The Federal Communications Commission (FCC) first undertook an examination of convergence in the 1966 Computer Inquiry.[18] The Commission issued Computer I in 1971, and decided to regulate telecommunications services and forbear from applying any Title II regulations contained in the Communications Act of 1934 to computer services. The FCC issued Computer II and III to respond to technological changes after Computer I. Computer II classified services into basic services, such as telephone services and enhanced services, to which computer processing was added, and decided to deregulate enhanced services. Although the term and service category were changed through the revision of the Computer inquiries, the regulatory demarcation between telecommunications and enhanced services was basically maintained. Refraining from regulation is generally seen as contributing to the promotion of competition, since deregulation enables carriers to take prompt and flexible action. However, the FCC was confronted with the difficult problem of classifying each service into either telecommunications services or enhanced services. The FCC has had to decide whether new services such as IP telephone and the Internet access service are telecommunications services or enhanced services.

In Japan, there was a move in the late 1960s to connect several mainframe computers with telecommunications networks. However, since NTT Public Corp. strictly limited the connection between computers and telecommunications networks in order to ensure the reliability of networks, no data processing service was provided at that time.[19] Taking into consideration the increase in demand and technological progress, MPT revised the Public Telecommunications Law in 1971 and authorized the provision of data processing services within strict limits. Then, although the MPT revised the law in 1982 at the request of users, computer services did not develop significantly, owing to the high prices.

The TBL, which took effect in April 1985, classified telecommunications carriers into Type I and Type II carriers. Type I carriers provided telecommunications services employing their own facilities, and Type II carriers provided their services by leasing facilities from Type I carriers. Type II carriers were able to provide any kind of service, but in fact their main service was data processing, from the standpoint of competition with Type I carriers.

Since the enforcement of the TBL in 1985, Japanese enhanced services have developed rapidly, owing to deregulation and competition. MPT adopted the classification of carriers by facility, instead of the classification by service that the FCC had adopted. Therefore, MPT was not confronted with significant problems regarding the demarcation of services.[20] This indicates that the appearance of intermediate services depends on the regulatory framework, including the definition of terms.

Compared with Type II carriers, Type I carriers were regulated more strictly, owing to the fact that network facilities constructed by Type I carriers were regarded as infrastructure. However, small-scale Type I carriers that provided their services within limited areas and large-scale Type II carriers that provided nationwide services appeared in the telecommunications market, and the difference between Type I and Type II carriers became less apparent. Therefore, MIC came to the conclusion that classification by facility was out of date and accordingly abolished Type I and Type II categories in April 2004. While the NTT East and NTT West companies have been regulated with regard to interconnection and the tariffs for basic telephone services in order to prevent the exercise of monopoly power in the local communications market, the streamlined regulations have been applied to other carriers.

Recently, Internet telephone services began to penetrate into both firms and households in Japan. Internet telephone is interactive voice communication over an Internet protocol, and consists of telephone terminals or personal computers and transmission facilities using an Internet protocol. In other words, Internet telephone is the combination of telecommunications and computers, and its interface is specified by the *de facto* standard. When a firm with a monopoly provided telecommunications services, a closed interface dominated the market, and the firm's revenues were guaranteed. However, modularity and a *de facto* standard now prevail in the computer industry, including the Internet, and have extended to the telecommunications industry. It could be said that the telecommunications industry is coinciding with the computer industry.

Selection of modularity with open interface and integration

The combination of modularity and an open interface has been applied to the telecommunications industry as well as to the computer industry, as already mentioned. As this combination enables telecommunications carriers to concentrate managerial resources in their business domain, the modularity system is well suited to this industry, in which technological progress occurs rapidly.

However, we have witnessed the integration of telecommunications carriers through mergers since the late 1990s. For example, the international telecommunications carrier KDD, the long-distance carrier DDI and the cellular phone provider IDO merged to become KDDI in 2000. As a result of the merger, KDDI became vertically integrated and provides a range of services from local communications to international communications. On the other hand, Softbank Corp. has established several subsidiaries and has merged with other companies, and provides a range of services such as POTS, ADSL, IP telephone and e-commerce. These are examples of integration and diversification. Since carriers with small-scale networks can provide telecommunications services worldwide by connecting with other carriers' networks under an open interface, integration through large-scale mergers and diversification seem to be contradictory in the telecommunications market. So, let us examine the selection of modularity with an open interface and integration.

Since the combination of modules and an open interface enables many newcomers to enter the market, price competition will be promoted. For example, Softbank Corp., a Type II carrier, leased telecommunications facilities from NTT and other Type I carriers, and expanded the area of ADSL services nationwide within a short period. Since Softbank Corp. significantly reduced the charges for ADSL services and acquired a large number of users, competitors were compelled to match the low rates that it set. Consequently, the rates for ADSL services have decreased, and the number of ADSL subscribers had reached 11 million as of March 2004. Since the number of broadband service subscribers has reached 15 million,[21] it can be said that ADSL is the driving force of the expansion of broadband services in Japan. In addition to the entry of newcomers, since services provided under the *de jure* and *de facto* standards are homogeneous, users view the services as substitutes for those provided by other carriers. Homogeneous services have intensified price competition. If Japanese regulators did not introduce unbundling that corresponded to the *de jure* standard, the prices of telecommunications services would have remained high.

On the other hand, integration of telecommunications carriers provides one-stop shopping and various kinds of tariffs. Several ISPs have provided an IP telephone service for the subscribers to their ADSL service without additional charges, with the purpose of acquiring ADSL subscribers. KDDI, which is engaged in cellular phone and fixed telephone services, has provided various bundled services, such as a discount on its international telephone service when international calls are made via a cellular phone. Discounting across a range of services gives users an incentive to purchase a bundle of services from an integrated carrier.

In addition, vertical integration has another advantage. For example, when ISPs distribute content through their websites, they have to negotiate with the content producers for copyright. While the procedure for copyright is generally complex and costly, producing content in-house reduces transaction costs relative to copyright. In addition, when ISPs provide attractive content on websites, they can enhance their brands and reputation. Therefore, integrated companies providing both transmission and content have competitive advantages. In fact, large-scale ISPs have recently become engaged in the production of content.

In conclusion, one purpose of mergers under an open interface is to respond to the price competition and acquire competitive advantages through the reduction of transaction costs, the enhancement of reputation and enclosure of users caused by demand-side factors. In addition to this, technological convergence between traditional telecommunications and computers has promoted a trend towards integration and alliances in the information industry.

Although vertical integration has several advantages, as mentioned, integrated companies do not always succeed. The merger of Time Warner and AOL is an example of vertical integration. AOL is a large ISP and Time Warner is a prominent content holder. As a result of the merger, AOL can distribute much of the content produced by Time Warner without the procedure of copyright, and attractive content is expected to enhance the value of the AOL website. However, some of the content distributed by AOL is procured from a third party, and not all content is produced in-house. Since broadband networks provide multiple channels and distribute a large amount of content, it is difficult for distributors to supply their channels completely with content that is produced entirely in-house. To take an example from broadcasting, CS broadcasting, which provides more than 100 channels, is vertically separated into transmission facilities and content produced by a third party, as mentioned earlier. In addition, Japanese terrestrial broadcasters own the transmission facilities and produce television programmes themselves in principle, according to the Broadcasting Law. However, since local stations do not possess adequate operational resources to produce all the programmes required to fill their television schedules, they receive programmes from large-scale broadcasters according to the network affiliate contracts. In addition, even large-scale terrestrial broadcasters often purchase television programmes from programme production companies for economic reasons. A television broadcaster provides only one analogue channel at present. With the expected increase in channels brought about by digitization,[22] it will be more difficult for a broadcaster to produce all programmes in-house, and the purchase of programmes from production companies will inevitably increase. Since producing a lot of attractive content needs managerial resources, distributors produce some content in-house and procure other content from content producers in general. In this respect, a company that merges with a content holder cannot entirely avoid copyright issues.

On the other hand, until the content is distributed in public, nobody knows whether or not the content produced in-house will be attractive to many viewers. If the content is attractive, the copyright holder gets additional revenue from secondary use of the content. For example, the film industry may repackage popular feature films and earn a significant amount in licensing fees from the sale of videos as well as cinema admissions. However, if a feature film is not attractive to a lot of viewers, the film is not used secondarily, and the film company may take a loss on the film. When distributors purchase content that has already been released publicly, they can select and procure popular content, based on how well the market has responded to that content. Thereby, they may reduce their risks by purchasing published content, although they need to pay copyright fees.

In the case of i-mode service provided by NTT DoCoMo, the cost of producing content for websites is relatively small, since the transmission capacity via a second-generation cellular phone is limited. Low production costs enable numerous content providers to enter the market,[23] in addition to the adoption of an open interface. The business model adopted by NTT DoCoMo presumes that production costs are low, and a lot of content producers will therefore enter the market. However, this presumption is not always applicable to all content distribution businesses. The production costs for content to fill channels are different among content via a second-generation cellular phone, television and feature films, and the selection of an appropriate production system depends on the cost of content production and the number of channels. Therefore, the differences in production cost and the number of channels lead to the differences in industry structures across industries. As a result, the two types of production process – that is, specialization and integration – can be observed in the information industry at the same time. Competition policy will be considered in the final part of this chapter.

Change of interface and its implications for competition policy

In this section, the changes in the interface and industry structure will be summarized and problems regarding competition policy will be identified. A few decades ago, the main determinant of industry structure for telecommunications was the cost structure. Market demand and cost function of firms controlled the number of firms in the market, and the existence of scale economies and scope economies controlled market performance. When the market was considered to be a natural monopoly, NTT Public Corp., as the state-owned public telephone operator, was the only service provider in Japan.

Since the introduction of competition, interconnection has been the main topic in Japanese telecommunications policy. Since subscriber lines have been regarded as essential facilities, MPT and the present MIC established the interconnection rules to promote competition. According to the rules, NTT's networks have been open in both a technical and an economic sense, and information disclosure has developed. In addition to the interconnection rules, the reorganization of NTT was conducted in order to bring about a level playing field.

To sum up, when Japanese telecommunications services were provided by a monopolistic firm, the firm was vertically integrated, and the network interface was closed information. Since April 1985, integrated networks have been divided into several modules, and the interface has been opened by the interconnection rules. We may say that interconnection rules have contributed to an improvement of transparency and information disclosure.

Some public telecommunications networks have recently been replaced by the Internet protocol network, which consists of modules and an open interface as a *de facto* standard. Since services produced under modularity with an open interface are homogeneous, firms seek a competitive advantage by utilizing demand-side factors such as one-stop shopping, diversified tariffs, network externalities and lock-in caused by switching costs in order to respond to price competition. Certainly the cost structure is one of the factors influencing firms and the industry structure at present. However, since a large amount of investment is not required in order to provide telecommunications services, especially IP services, the importance of cost structure as a determinant of industry structure and competitive advantage has been reduced. On the other hand, since networks have been opened up with the development of the Internet and the rules, demand-side factors have become more important in the acquisition of a competitive advantage.

Recently, mergers between large-scale telecommunications carriers have occurred under an open interface. Diversified companies are in a favourable position to enclose their users by employing demand-side factors such as one-stop shopping and network externalities. In addition, telecommunications carriers other than NTT East and NTT West are able to set their rates strategically, owing to the deregulation. Therefore, it is possible that integrated companies will dominate the market, even if their networks consist of modules and an open interface. Although interconnection has been discussed and prescribed by law, to date, few studies on competition policy have incorporated the viewpoint of the demand side, and it is difficult to reach a consensus on the matter, as can be seen from the antitrust action brought against Microsoft.[24] Competition policy under an open interface is an issue that requires further consideration.

Regulations are designed on the basis of a review of the competitive environment. For a few years after the introduction of competition, the market share and the number of newcomers entering the telecommunications market were major factors that had to be grasped in order to gauge the progress of competition in the market. The data relating to the market share and the number of carriers are certainly objective and essential information, but are not sufficient to evaluate competition. Although the number of telecommunications carriers exceeded 12,500 at the end of March 2004, the market has actually been dominated by a few large-scale carriers. Although at present even large-scale carriers cannot maintain high profit margins, owing to price competition and the threat of a potential competitor using new technology, it is not certain whether the current competitive mechanism will work in the future. It is important to monitor the price–cost margins of carriers that have a competitive edge with regard to prices and anticompetitive conduct such as tie-in sales, although collecting information on the behaviour of carriers is not easy. In addition, we should note that integration tends to reduce transparency in business activities such as in decisions on prices.

Finally, the problems regarding competitive policy that we need to consider will be identified. The business domains of the NTT East and NTT West Companies have been limited to the local communications market, and their entry into the long-distance communications and broadcasting markets has been forbidden in order to avoid undue exercise of market power. Therefore, while other carriers have determined their business domains based on their management strategies, and their capacity to act flexibly has been guaranteed as a result of the streamlined regulations, NTT East and NTT West have been regulated with regard to line of business as well as interconnection and the rates charged for basic telephone services. This may lead the two NTT companies to compete with their rivals under disadvantageous conditions. Although asymmetric regulation is regarded as one means of deregulation in the sense that the subjects of regulation are limited, one cannot rule out the possibility that an arbitrary design of asymmetric regulation will distort the competitive conditions in the market.

In addition, not only metal subscriber lines but also optical fibre subscriber lines are the subjects of unbundling in Japan.[25] Therefore, NCCs that do not own their broadband local networks can provide broadband services by leasing optical fibre local networks from the two NTT local companies. This may mean that both NTT and other carriers have no incentive to deploy optical fibre local networks, since the demand for broadband services is uncertain and the cost of deploying networks is large, especially in rural areas. The definition of essential facility is not fixed, and is expected to change along with technological progress. Incentives for investment should be considered while policymakers formulate the interconnection rules and other regulations.

Notes

1 For path dependence and institutional diversity, see Masahiko Aoki and Masahiro Okuno (eds), *Keizai System no Hikaku Seido Bunseki* (A Comparative Institutional Analysis of Economic Systems) (Tokyo: University of Tokyo Press, 1996).

2 For an empirical test of the positive feedback mechanism that NTT DoCoMo has enjoyed, see Tatsuo Tanaka, 'Keitai Denwa Sangyo niokeru Network Gaibusei no Jisho' (Econometric Evidence of Network Externalities of the Mobile Phone Industry in Japan), *Mita Gakkai Zasshi* (Mita Journal of Economics) 95 (3): 119–132 (2002).

3 See also Carliss Y. Baldwin and Kim B. Clark, 'Managing in an Age of Modularity', *Harvard Business Review*, September/October: 84–93 (1997).

4 See Carliss Y. Baldwin and Kim B. Clark, *Design Rules: The Power of Modularity* (Cambridge, MA: MIT Press, 2000).

5 See Takahiro Fujimoto, Akira Takeishi and Yaichi Aoshima (eds), *Business Architecture* (in Japanese) (Tokyo: Yuhikaku, 2001).

6 See Sumiko Asai, *Zyoho Sangyo no Togo to Mozyuruka* (Integration and Modularity in the Information Industry) (Tokyo: Nippon Hyoron, 2004).

7 The ratio of NTT procurements to total sales of the four manufacturers exceeded 0.4 during the period 1979–1983. For this figure, see Teruyuki Inoue, *Denden Mineika Katei no Kenkyu* (Studies on the Process of Privatization of NTT Public Corp.), table 10-2, pp. 308–311 (Tokyo: Eruko, 2000).

8 Since the cost of collecting interconnection charges from NCCs is lower than

the cost of collecting charges from a lot of users, interconnection charges should in principle be lower than users' charges.

9 According to the agreements, the sectors of mobile services and data processing services were separated from NTT, and NTT Data and NTT DoCoMo were established in 1988 and 1992, respectively.

10 For the number of carriers, see the White Paper on Information and Communications in Japan in 2004, pp. 140–141. It is available at http://www.soumu.go.jp/.

11 In the case of fixed telephone services, users can change their carriers without having to change telephone numbers, and number portability has been guaranteed. In the case of cellular phone services, number portability will be introduced in 2006 in Japan. However, regarding email addresses, portability will not be guaranteed. For the number portability of cellular phone services, see the MIC press release dated 27 April 2004, available at http://www.soumu.go.jp/.

12 For example, KDDI, a large-scale Japanese telephone company, issued a statement that the introduction of an IP-based system to replace the existing fixed-line phone network will begin in fiscal year 2005 and the replacement using software for switching will be completed by the end of March 2008. See the press release from KDDI dated 15 September 2004, available at http://www.kddi.com/.

13 Since the reorganization of NTT in July 1999, the NTT East and West companies have provided *Dial* Q^2 services.

14 However, the chat service was suspended in 1994, since social problems such as obscene and indecent conversations occurred.

15 While telecommunications carriers have been regulated by the Telecommunications Business Law, broadcasters have been regulated by the Broadcasting Law.

16 The guidelines are available at http://www.mpt.go.jp/ and http://www.soumu.go.jp/.

17 For data on the number of BS and CS subscribers, see the White Paper on Information and Communications in Japan in 2004, p. 172, available at http://www/soumu.go.p/.

18 For the Computer Inquiry in detail, see Peter W. Huber, Michael K. Kellogg and John Thorne, *Federal Telecommunications Law*, 2nd edition, pp. 1089–1095 (New York: Aspen Publishers).

19 In Japan, the term 'data processing service' has been generally used, instead of the term 'enhanced service'.

20 However, in some cases MPT had to coordinate with the Ministry of International Trade and Industry (MITI, the present Ministry of Economy, Trade and Industry) since MITI had jurisdiction over the data processing services.

21 For the number of broadband subscribers, see the White Paper on Information and Communications in Japan 2004, p. 4, available at http://www.soumu.go.jp/.

22 One analogue channel corresponds to three digital channels, owing to the efficient use of frequencies, as long as broadcasters do not provide high-definition television (HDTV).

23 The numbers of i-mode official sites and non-official sites are 4,432 and 81,153 respectively, as of November 2004. See http://www.nttdocomo.co.jp/.

24 One of the factors in Microsoft's market power is considered to be the existence of network externalities resulting from the penetration of Windows and the complementarities between OS and applications software programs.

25 In general, local telephone carriers in the United States do not have the obligation of unbundling in relation to optical fibre subscriber lines. The FCC decided in October 2004 to limit the unbundling obligations imposed on FTTC (fibre-to-the-curb) deployments as well as FTTH (fibre-to-the-home) deployments to remove disincentives to the deployment of advanced telecommunications facilities. For the detailed FCC decisions, see FCC Decision and Reconsideration, FCC 04-248, available at http://www.fcc.gov/.

2 The broadband market in Japan

Takanori Ida

Introduction

Recall that only a few years ago we were convinced that Japan's telecommunications industry lagged behind those of Europe and the United States by some ten years. However, the recent development of broadband services in Japan has been remarkable. According to the Ministry of Internal Affairs and Communications (MIC), as of August 2004 the penetration rate of broadband services had reached 34 per cent of Japanese households, broken down as follows: (1) asymmetric digital subscriber lines (ADSL), 74.2 per cent; (2) cable television (CATV) Internet, 16.4 per cent, and (3) fibre to the home (FTTH) 9.5 per cent. Yahoo! BB, owned by Softbank Co. Ltd, currently maintains the top spot in the ADSL market, while K-Opticom of the Kansai Electric Power group is competing fiercely with NTT West for the FTTH market. At the same time, the low-speed ADSL market has been absorbed by the medium-speed ADSL market, whose users are now converting to the high-speed ADSL and FTTH markets.

Given such competitive circumstances in Japan's broadband markets, in June 2004 MIC published *The Effective Competition Review of Japan's Telecommunications*, which examined Internet access services for fiscal year (FY) 2003. In order to put the regulatory reform into practice, the following four points should be stressed: first, to understand how plain old telephone services (POTS) are substituted by IP-based services; second, to predict what information and communications technology (ICT) will spread in the future; third, to ensure the transparency of the competition policy in the Japanese telecommunications industry; and fourth, to coordinate the international developments of telecommunications policies. We accordingly collaborated with MIC to delineate the relevant markets and analysed the substitutability of various forms of Internet access services, based on econometric analysis such as nested logit model and conjoint analysis. The purpose of this chapter is, first, to summarize *The Review*, and second, to introduce our empirical research. We shall avoid mathematical details and rely on the intuitive ability of readers to understand graphs. For the moment, we exhibit two important conclusions of this chapter.

First, using a nested logit model based on that of Ida and Kuroda (2004), we analysed the access demand to such narrowband Internet services as dial-up (DU) and integrated services digital network (ISDN) as well as such broadband Internet services as ADSL, CATV Internet, and FTTH. The nested logit model can flexibly demonstrate multi-level choice structures. It divides the choice set into categories: individuals choose one category among divided categories (including narrowband and broadband services) and then determine a specific choice (including ADSL, CATV-Internet and FTTH services) from the chosen category. The nested logit model relaxes the 'independence from irrelevant alternatives' (IIA) property that the ratio of choice probabilities is determined only by the relevant two alternatives and is completely independent of any other alternatives. After measuring the own-price elasticities of access demand, represented by modulus, only the ADSL figure is inelastic (0.29), while the CATV and FTTH figures are on the borderline between elastic and inelastic (0.88~1.11). Thus, we conclude that the ADSL market is independent of other services. Furthermore, since the present Japanese broadband market is overwhelmingly dominated by ADSL, we divide it into three submarkets: low-speed, medium-speed and high-speed. After measuring the own-price elasticities, the medium-speed ADSL figure is very inelastic (0.15) but, interestingly, the low- and high-speed ADSL figures are highly elastic (7.32 and 6.74). The periphery of the large ADSL market is actually competing on both sides with narrowband services (DU and ISDN) and broadband services (CATV and FTTH).

Second, we demonstrate the results of conjoint analysis for both the users who can and those who cannot avail themselves of FTTH services, based on the work of Ida and Sato (2004). Conjoint analysis aims to measure consumers' stated preferences (SP) based not on the actual data observed in the market, but on the answers to a virtual questionnaire, which highlights a remarkable difference from the revealed preference (RP) method. One advantage of conjoint analysis is that it is virtual by nature and scrutinizes consumers' preferences that are generally ignored or overlooked. For instance, even when we cannot collect actual data concerning a product prior to its market launch, we can quantitatively evaluate such products. Consequently, we recognize that stated preference (SP) depends on the actual availability of FTTH; willingness to pay (WTP) for 1 Mbps is about ¥30 ($0.27, given ¥1 = $110) for people with access to FTTH, while it is about ¥70 ($0.63) for people without access. Next, we compare SP and revealed preference (RP) in identical populations with access to FTTH. As a result, we see that their SP and RP are different; the WTP for 1 Mbps is about ¥30 ($0.27) based on SP, while it is about ¥20 ($0.18) based on RP.

This chapter is composed of five sections. The first provides preliminary discussions that consider the success of Japan's broadband services. The second summarizes *The Effective Competition Review of Japan's Telecommunications*, published by MIC in June 2004. The section after that introduces research by Ida and Kuroda that analyses the demand substitutability of broadband services, which is followed by a section that introduces research by Ida and Sato that applies conjoint analysis to consumer preferences for broadband services.[1] The final section discusses problems to be solved by Japan's broadband market in the future.

Why success in Japan?

The current development of Japan's broadband Internet access services is remarkable. Figure 2.1 shows how the number of Internet users has changed. The number of narrowband Internet users has decreased, but that of broadband Internet users has increased. In August 2004 the latter overtook the former. At present, the penetration rate of Japan's broadband services is around 34 per cent.

Next, we turn to a breakdown of broadband services. Figure 2.2 illustrates how the number of broadband Internet users has changed. First, we see that ADSL has played a leading role in the development of Japan's broadband services. The number of ADSL users reached ten million in 2003, ranking Japan with Korea as a major ADSL-using nation. Second, the number of FTTH users continues to grow steadily, having exceeded one million in 2004. At the moment, Japan has no peer in FTTH diffusion. Third, even though the CATV Internet user market is more than two million strong, its growth rate is small. In this respect, the Japanese broadband market is different from that of United States, where CATV Internet is the most popular.

Today, Japan's broadband service is reputedly the cheapest and the fastest in the world. An International Telecommunication Union (ITU) Internet Report entitled *The Birth of Broadband* compared rates per 100 kbps among

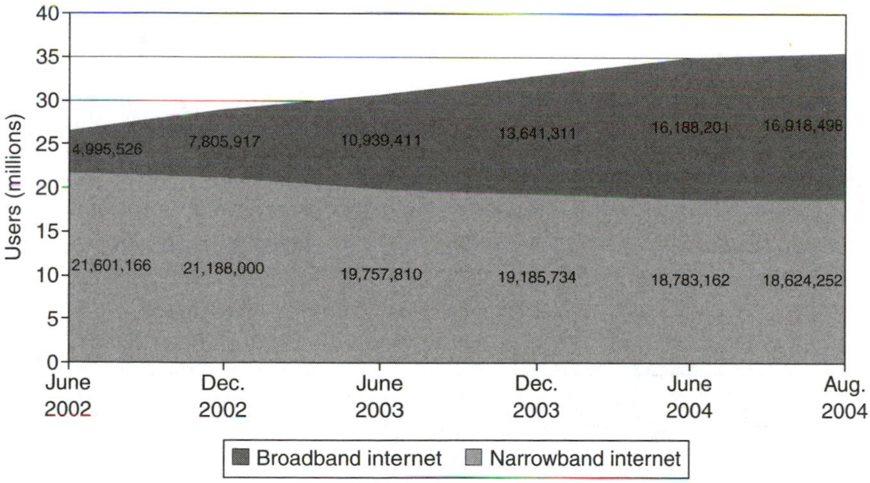

Figure 2.1 Changes in the number of Internet users (source: http://www.soumu.go.jp/w-news/2004/040930_2.html).

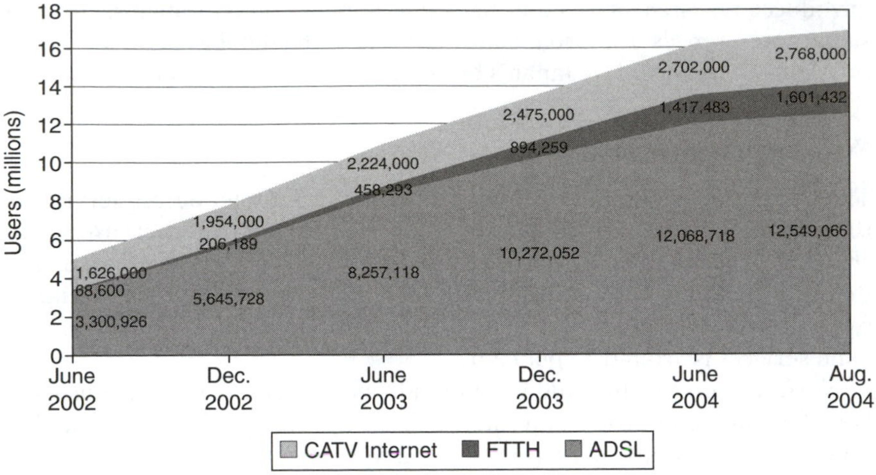

Figure 2.2 Changes in the number of broadband Internet users (source: http://www.soumu.go.jp/w-news/2004/040930_2.html).

various countries as of September 2003. Figure 2.3 shows that Japan ($0.09) is much cheaper than the United States ($3.53) and elsewhere.

It has been said, however, that in Japan, telecommunications lags ten years behind Europe and the United States. Why are their positions now reversed? At this point, two hypotheses explain this fact.

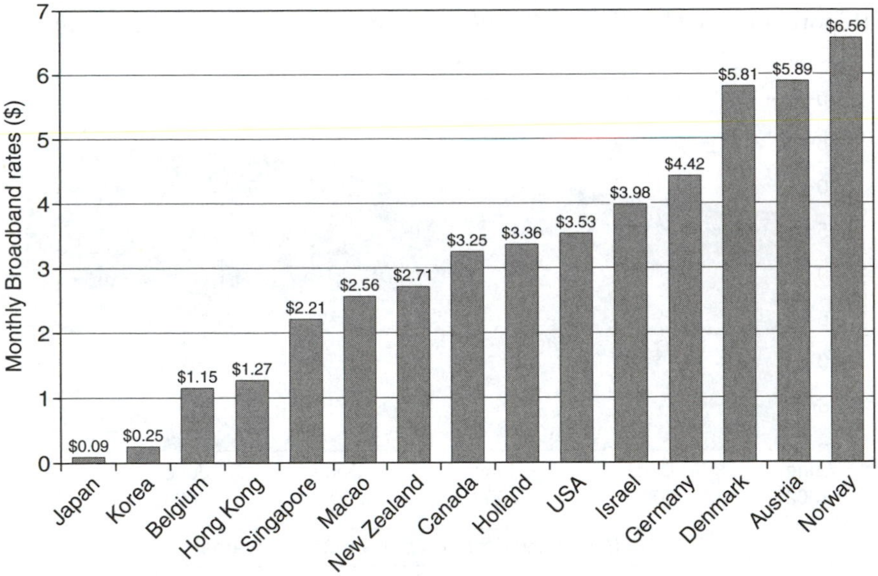

Figure 2.3 International comparison of broadband monthly rates per 100 kbps (source: ITU Internet Report 2003: Birth of Broadband, 2003.9).

First, Japan's telecommunications policies have produced two results, one that met expectations and another that was unexpected. The latter success is based on the fact that the NTT group was not broken up. In 1990 and 1996, the Ministry of Posts and Telecommunications (the present Ministry of Internal Affairs and Communications) submitted reports that requested the divestiture of the NTT group; but after the Japanese government experienced difficulty dissolving it, the company was finally reorganized under a holding company system. NTT has long supported a Broadband Integrated Services Digital Network (B-ISDN) project in which NTT would establish fibre-optic access networks for all residences by 2015. Therefore, when the divestiture decision was overturned, the NTT group continued to invest into the construction of a broadband network. On the other hand, the expected success is based on a through open access policy of NTT's regional telecommunications network by MIC. In 1994, a cost-based interconnection charge system was introduced; in 1996, interconnection charges consisted of a call set-up fee plus a per second charge, with the opening of local networks; in 2000, a long-run incremental cost rule was adopted. Accordingly, in this industry Japanese transparent and radical open network policies have greatly promoted effective competition.

Second, many distinctive entrants have conveniently emerged while NTT was carrying out corporate streamlining – reluctantly at first, but genuinely afterwards. Since scale economies and network externalities apply to the information and telecommunications industries, we cannot expect perfect competition. Therefore, new entrants that have both the capability and the willpower to compete with the incumbents are necessary to maintain effective competition in the market. Fortunately, in the Japanese telecommunications industry many innovative entrants have emerged: DDI of the Kyocera group entered the long-distance telephone market, Yahoo BB! of the Softbank group entered the ADSL market, K-Opticom of the Kansai Electric Power Company joined the FTTH market, and so on. On the other hand, NTT confronted its competitors and reorganized under a holding company system in 1999 and drastically restructured its business in 2002. Consequently, stable and efficient competition has been realized in the Japanese broadband market.

Review of Japan's broadband market

As shown in the previous section, in Japan's broadband services the ADSL market led the way, followed by the FTTH market. Competition over price and for market shares is very fierce. In June 2004, MIC published *The Effective Competition Review of Japan's Telecommunications* (MIC, 2004 hereafter), which placed special emphasis on the broadband Internet access market. My purpose here is to introduce *The Review* and to explain the present situations of the ADSL, FTTH, and CATV-Internet markets.

Review of the ADSL market

ADSL users account for more than 70 per cent of Japan's total broadband market. To begin with, we consider the scale of the ADSL market. Figure 2.4 shows changes in the number of ADSL users. The figure has steadily increased, reaching a household penetration rate of 20.9 per cent by the end of 2003. By prefecture, Tokyo has the highest penetration rate (30.3 per cent), while Kagoshima has the lowest (8.2 per cent). Regional disparities in the diffusion of ADSL remain fairly large.

Figure 2.5 shows changes in the increasing rates of ADSL users. After the explosion of ADSL services, the rate of increase gradually decreased. From the end of 2002 to the end of 2003, the figure decreased from 30 per cent to 10 per cent because of rapid changeovers from ADSL to FTTH.

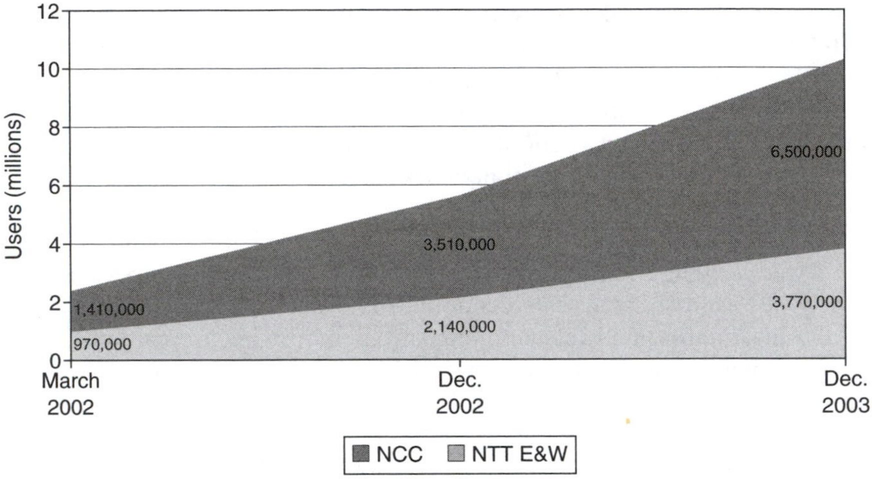

Figure 2.4 Changes in the number of ADSL users (source: MIC, 2004, Fig. A-1-3, p. 147).

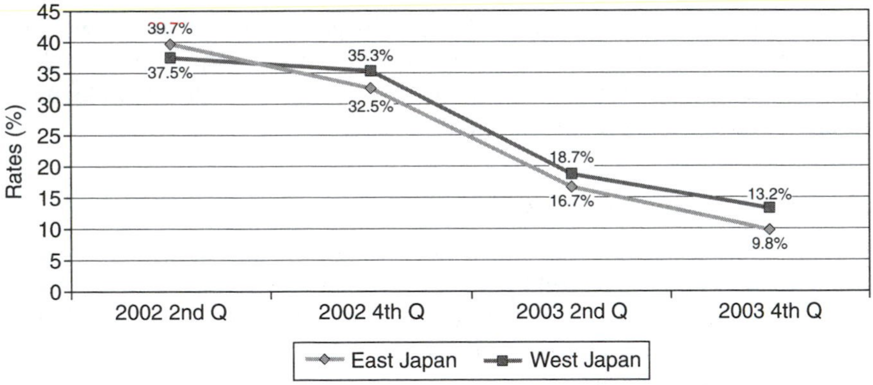

Figure 2.5 Trends in the increasing rates of ADSL users (source: MIC, 2004, Fig. A-1-10, 11, p. 154).

Second, let us consider market shares. Figure 2.6 shows the transition of ADSL market shares. NTT East and NTT West, the top two ADSL providers, have a 37 per cent market share in eastern and Western Japan, respectively. Note, however, that NTT East's market share has constantly remained a little over 35 per cent while NTT West's has suddenly decreased from the 40 per cent level. The competition that started in the Tokyo area has finally reached Western Japan.

Third, let us consider ADSL market concentration. Figure 2.7 shows the Herfindahl–Hirshman Index (HHI). The figure exceeds 2,500, implying strong oligopolistic tendencies. If we scrutinize the details, we see that concentration decreased up to September 2002, reflecting the decline

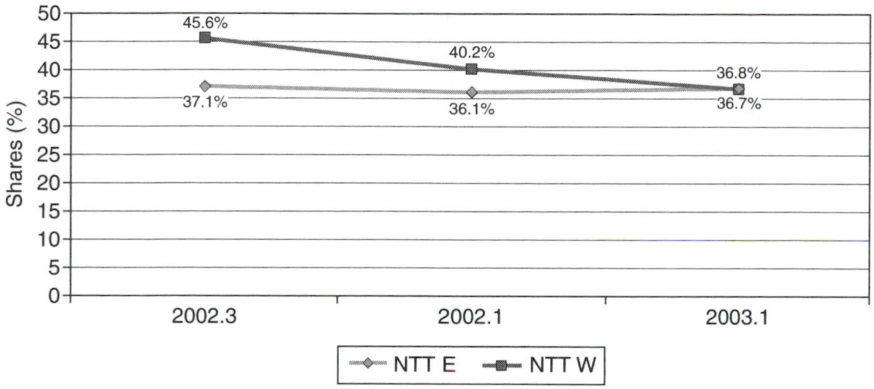

Figure 2.6 Changes in ADSL market share (source: MIC, 2004, Fig. A-1-10, 11, p. 154).

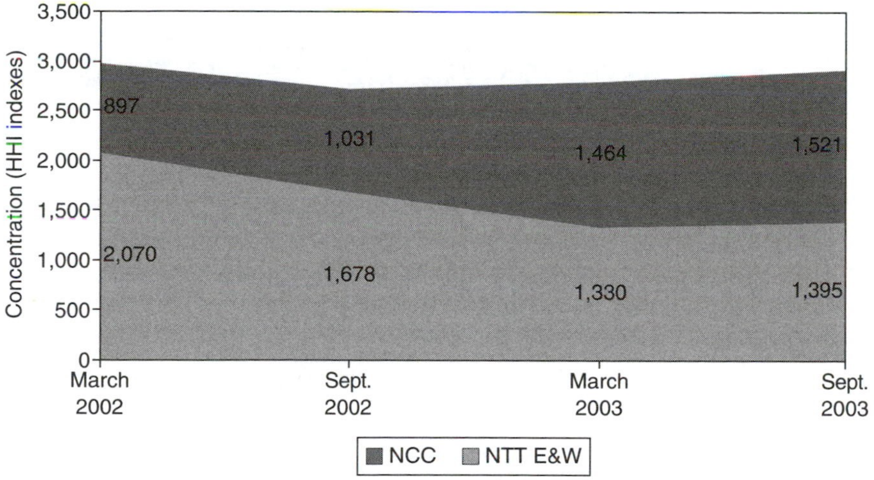

Figure 2.7 Changes in ADSL market concentration (Hirfindahl–Hirshman Index) (source: MIC, 2004, Fig. A-1-23, p. 166).

in the market shares of NTT East and West, but then an upward trend began that reflected the increase in Softbank's market shares. Figure 2.8 illustrates the transition of the ADSL market shares for the top three companies. The figure is larger than 80 per cent and still rising.

Fourth, let us address the level of ADSL rates, as shown in Figure 2.9. Such incumbent ADSL providers as NTT East, NTT West and eAccess reduced their rates, largely in response to Softbank's surprising entry in 2001. Figure 2.10 compares international ADSL rates. Alongside Korea, Japan has the cheapest rates per 1 kbps.

Review of the FTTH market

FTTH users comprise about 10 per cent of Japan's broadband market. First, let us look at the scale of the FTTH market. Figure 2.11 shows the changes in the number of FTTH users. The figure is rapidly increasing. Figure 2.12 illustrates the transition of the increasing rates of FTTH users. The national average is above 40 per cent; furthermore, NTT East and NTT West are growing faster than the national average.

Second, we turn to the FTTH market shares. Figure 2.13 shows the transition of the FTTH market shares. Note that NTT East and West together account for about 60 per cent of the total market. The growth of NTT West is especially remarkable. Among power-affiliated companies, K-Opticom of the Kansai Electric Power group enjoys a large market share. Since NTT West and K-Opticom are competing fiercely, Japan's broadband market demonstrates a *west-high/east-low* tendency: high penetration in Western Japan and low penetration in eastern Japan.

Third, let us consider FTTH market concentrations. Figure 2.14 shows

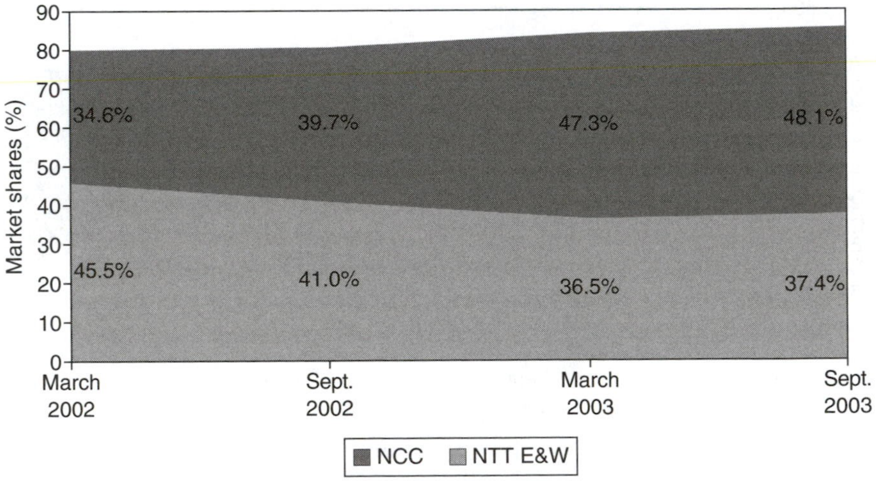

Figure 2.8 Changes in ADSL market shares of the top three (source: MIC, 2004, Fig. A-1-24, p. 167).

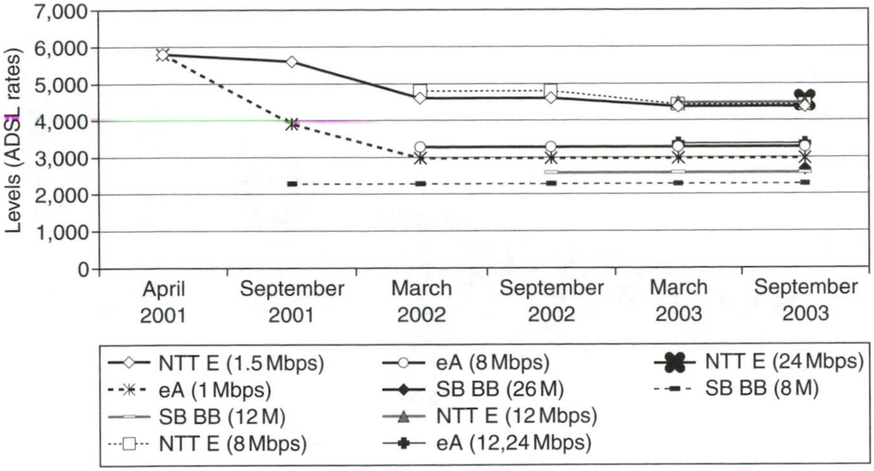

Figure 2.9 Changes in levels of ADSL monthly rates (source: MIC, 2004, Fig. A-1-29, p. 172).

Note
Figures exclude modem rental charges, and ISP is Biglobe.

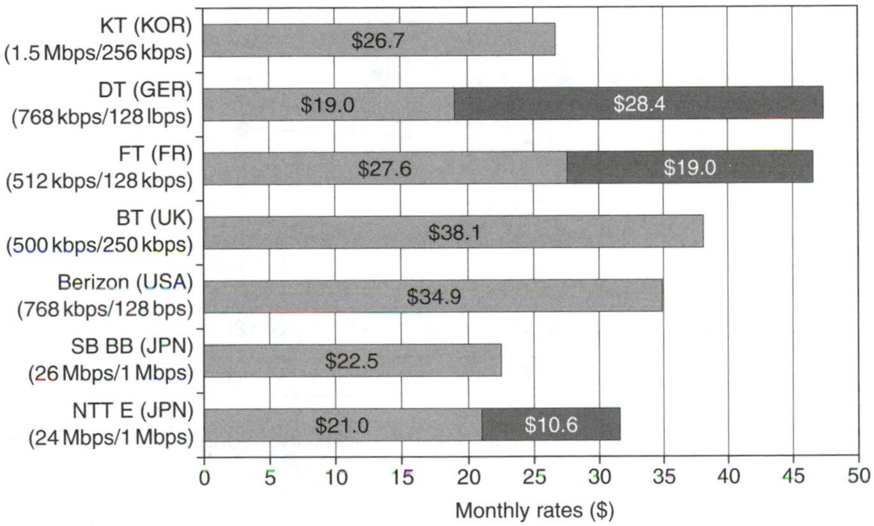

Figure 2.10 International comparison of ADSL monthly rates (source: MIC, 2004, Fig. A-1-30, p. 173).

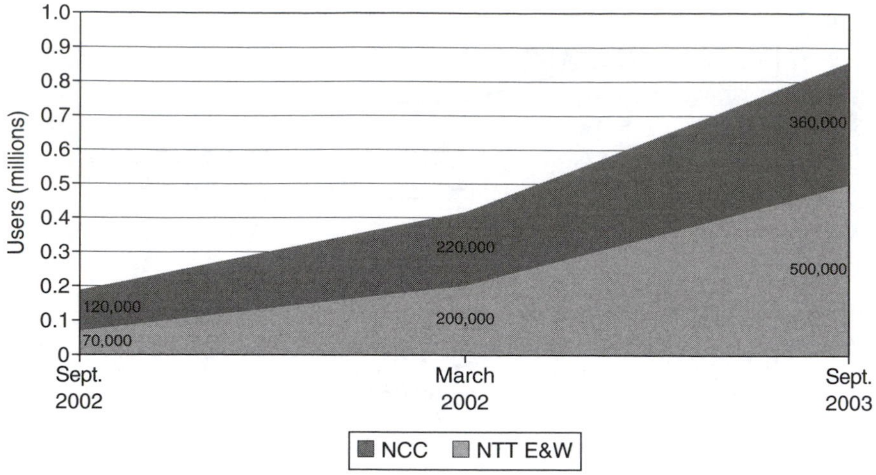

Figure 2.11 Changes in the number of FTTH (fibre-to-the-home) users (source: MIC, 2004, Fig. F-1-1, p. 216).

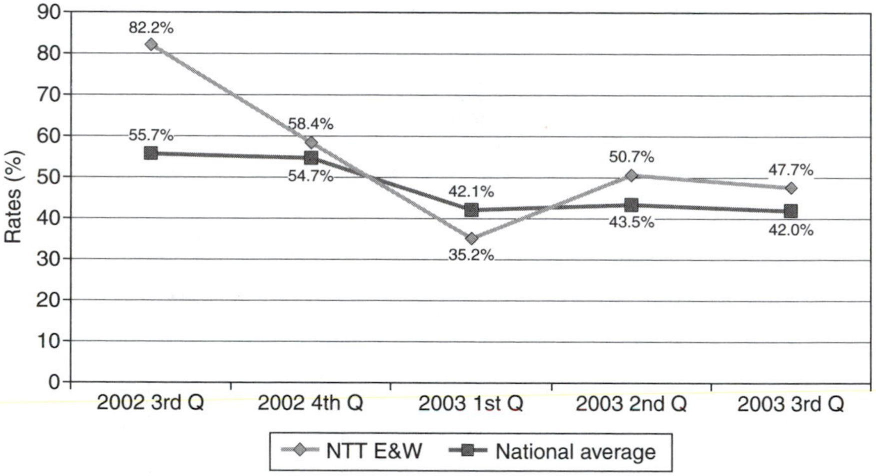

Figure 2.12 Trends in the increasing rates of FTTH use (source: MIC, 2004, Fig. F-1-1, p. 216).

the transition of HHI Indexes in the FTTH market. The figure has risen sharply to 3,000, creating an extremely oligopolistic market that reflects the increase in the market shares of NTT East and NTT West. Figure 2.15 illustrates the transition of the FTTH market shares of the top three companies. The figure has reached 80 per cent. However, it is necessary to divide the FTTH market into two submarkets: houses and apartments. As shown in Figure 2.16, NTT East and NTT West dominate rivals in the

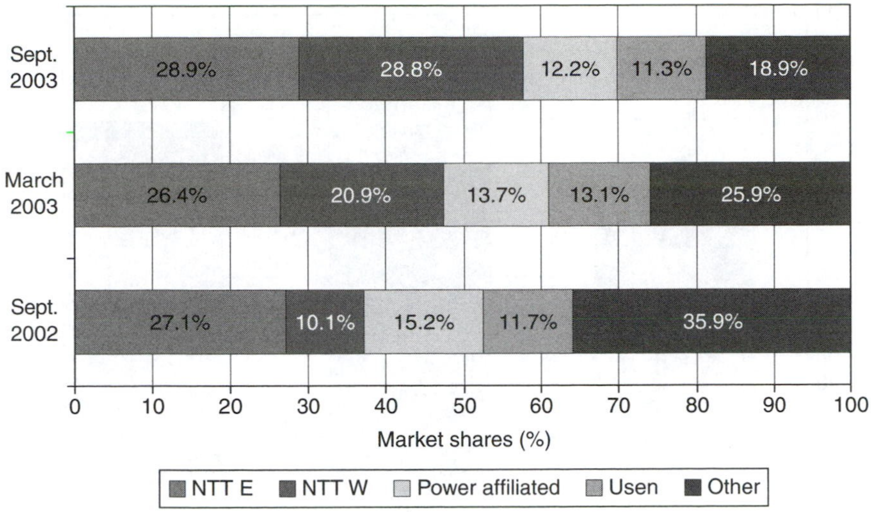

Figure 2.13 Changes in FTTH market shares (source: MIC, 2004, Fig. F-1-6, p. 221).

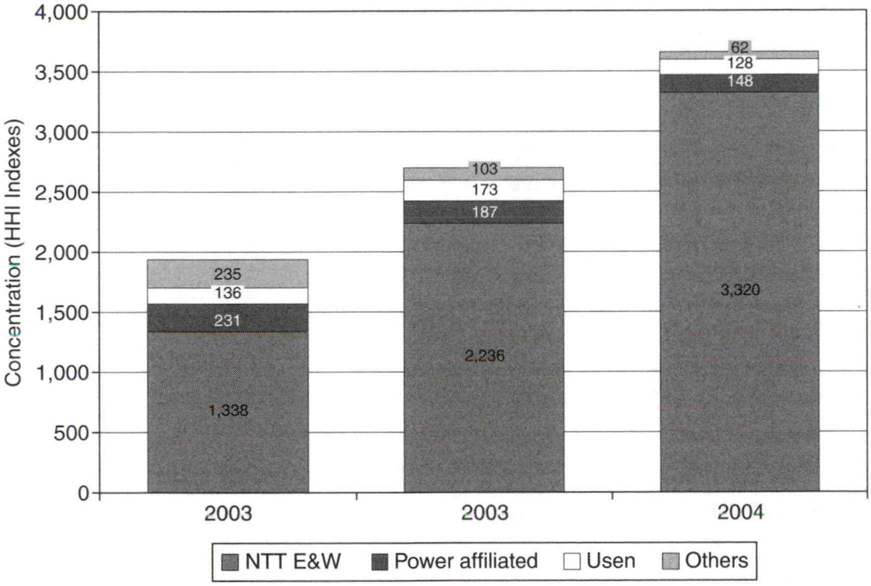

Figure 2.14 Changes in FTTH market concentration (Hirfindahl–Hirshman Index) (source: MIC, 2004, Fig. F-1-6, p. 221).

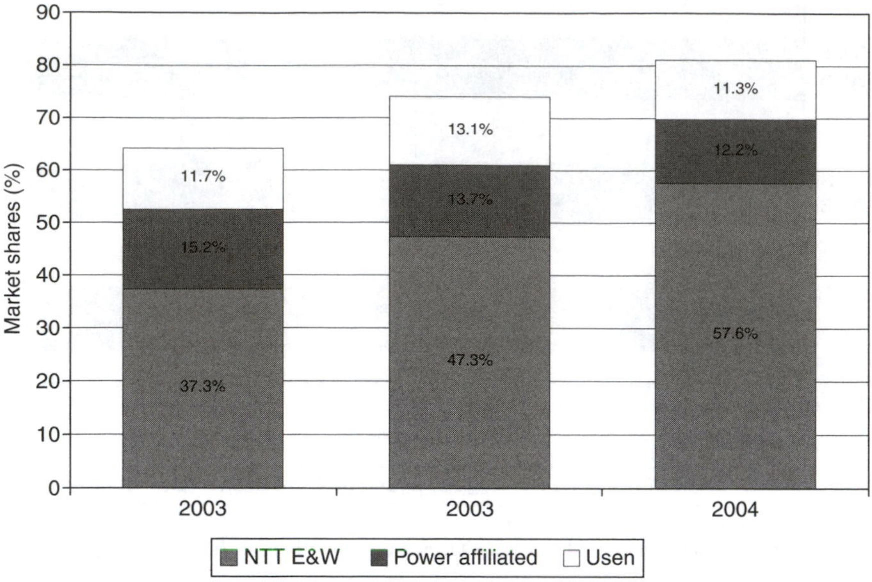

Figure 2.15 Changes in the FTTH market shares of the top three companies (source: MIC, 2004, Fig. F-1-10, p. 225).

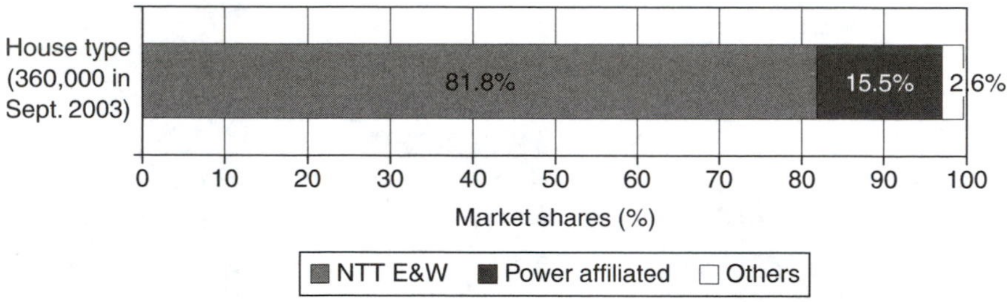

Figure 2.16 FTTH market shares, houses, September 2003 (source: MIC, 2004, Fig. F-1-10, p. 225).

house market. On the other hand, as shown in Figure 2.17, NTT East, NTT West and USEN are almost equally matched in the apartment market.

Fourth, what about the level of FTTH rates? Figure 2.18 shows the trend in FTTH rates. Since the beginning of the service, each company has lost levels. Even though the FTTH market is highly oligopolistic, the difference between ADSL and FTTH rates is so small that price competition across markets seems to be working. In the future, since we expect a rapid changeover from ADSL to FTTH, Japan will set an interesting precedent in which the migration of broadband services becomes problematic around the world.

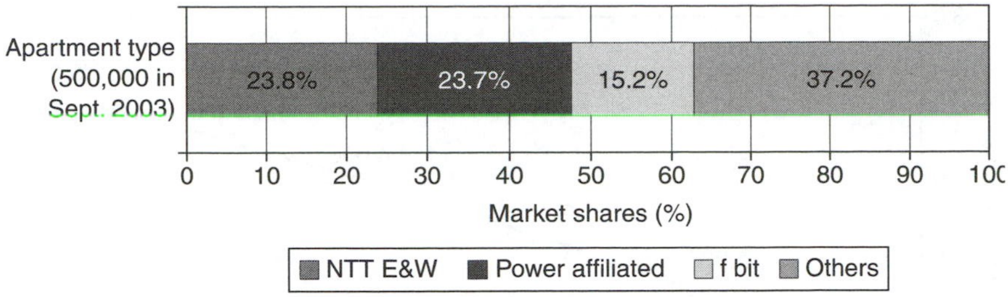

Figure 2.17 FTTH market shares, apartments, September 2003 (source: MIC, 2004, Fig. F-2-8, 9, p. 245).

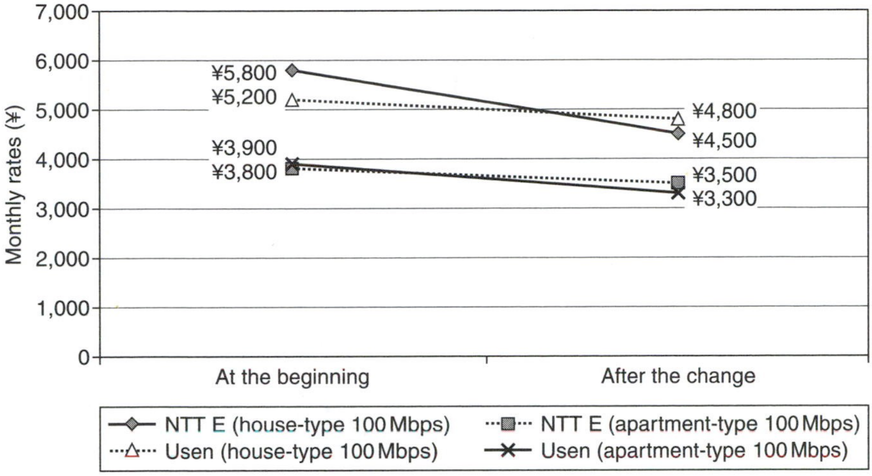

Figure 2.18 Changes in FTTH monthly rates (source: MIC, 2004, Fig. F-1-1, p. 226).

Review of the CATV Internet market

CATV Internet users comprise about 20 per cent of Japan's broadband market. We first look at the scale of the CATV Internet market. Figure 2.19 shows changes in the number of CATV Internet users, which is steadily increasing. Figure 2.20 illustrates how the number of CATV Internet providers has changed. Since the number is larger than 300, the market looks competitive. Figure 2.21 compares international CATV Internet tariff. We conclude that Japan has the cheapest CATV Internet rates. However, note that CATV Internet plays not a leading but an important supporting role in Japan's broadband market.

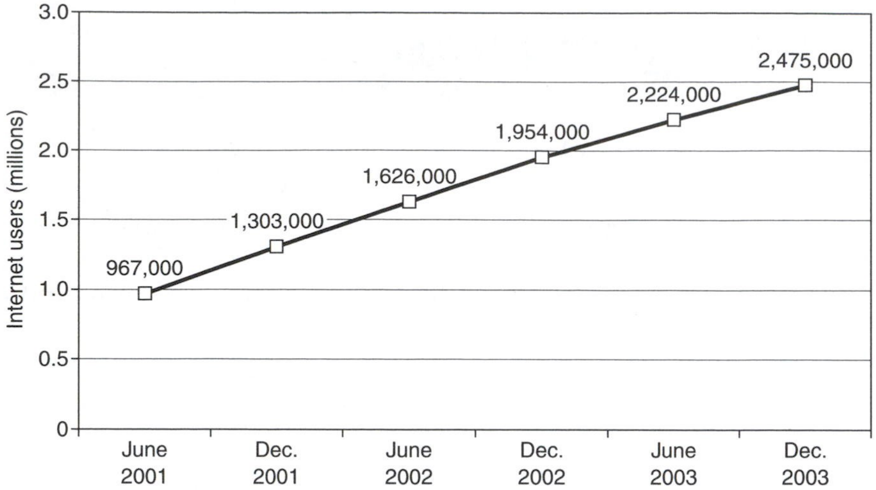

Figure 2.19 Changes in the number of cable television Internet users (source: MIC, 2004, Fig. C-1-1, p. 377).

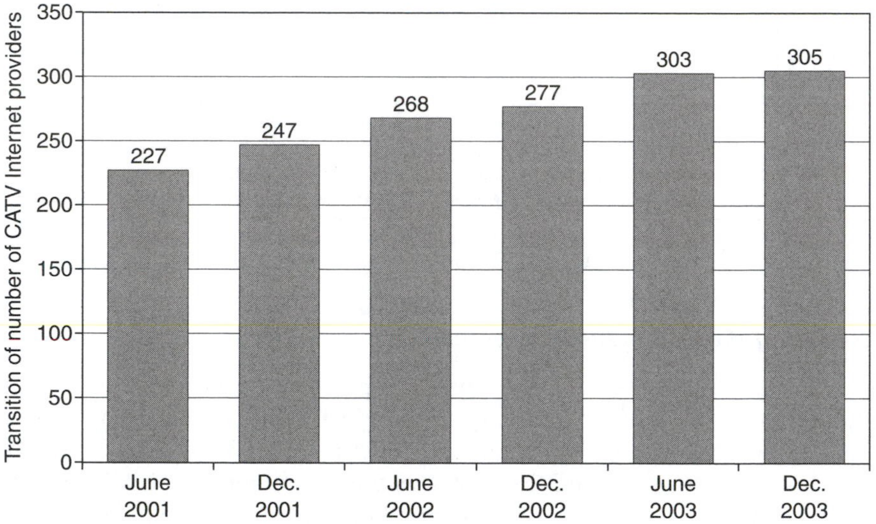

Figure 2.20 Changes in the number of cable television Internet providers (source: MIC, 2004, Fig. C-1-1, p. 377).

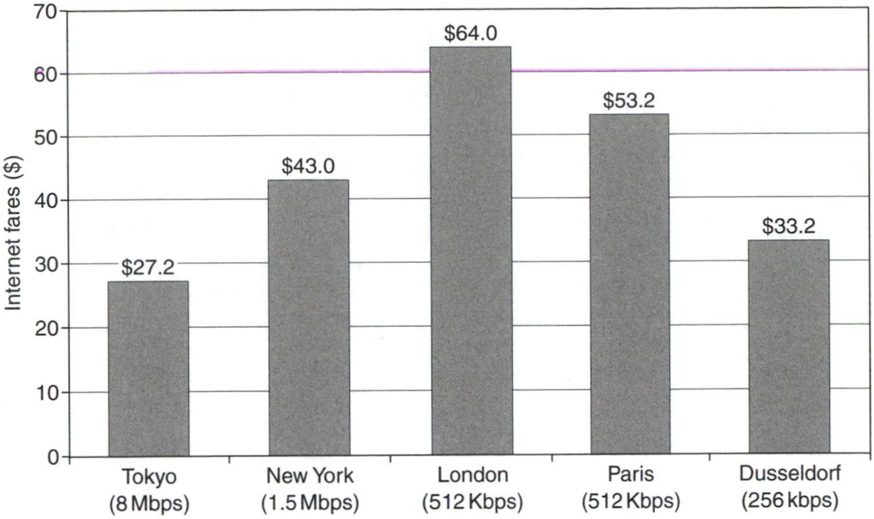

Figure 2.21 International comparison of cable television Internet tariffs (source: MIC, 2004, Fig. C-1-1, p. 377).

Demand substitutability of broadband services

This section investigates demand substitutability based on Ida and Kuroda's study. First, we explain the data and the descriptive statistics, and then we discuss the estimation results for broadband demand using a discrete choice model called a nested logit model. Considering the own-price elasticities of access demand, we found that ADSL is very elastic while FTTH and CATV are borderline cases.

Data and descriptive statistics

In the following, we explain the descriptive statistics and the data collected by a series of surveys of Internet access demand for private use conducted according to *The Guidelines for the Competition Review of Japan's Telecommunications* and *The Implementation Manual for FY 2003*, published in November 2003 by MIC.

The survey was carried out as a Web questionnaire, and a representative sample was randomly chosen from households with access to all five Internet alternatives: (i) dial-up (DU); (ii) always-on ISDN; (iii) ADSL; (iv) FTTH; and (v) CATV Internet. The total number of observations was 1,013; excluding omissions and abnormalities, we obtained 799 observations for nominal speed data. Questions included: (i) average expenditures per month (price); (ii) nominal access speed; (iii) type of Internet access and service provider; and (iv) such individual characteristics as gender, age, income, type of residence, living area, occupation, and so on.

The basic descriptive statistics are shown in Figure 2.22. The selection ratios were DU (2 per cent), ISDN (5 per cent), ADSL (67 per cent), FTTH (8 per cent) and CATV (18 per cent). Since the number of DU and ISDN users was very limited, it was difficult to consider them independent alternatives; therefore, we combined DU and ISDN into one narrowband alternative. Average monthly expenditures, defined as the sum of connection fees and ISP charges (and communication charges for dial-up Internet users), are shown as follows: DU: ¥3,946 ($35.90), ISDN: ¥5,207 ($47.30), ADSL: ¥4,344 ($39.50), FTTH: ¥5,929 ($54.10) and CATV: ¥5,200 ($47.30). Narrowband services are not always cheaper than broadband services because their charges are usage sensitive while those for broadband are flat. Among broadband services, FTTH is the most expensive while ADSL is the cheapest, as obviously expected. Average nominal access speeds are shown as follows: DU: 52 kbps; ISDN: 65 kbps; ADSL: 10 Mbps; FTTH: 82 Mbps; and CATV: 11 Mbps. A huge gap exists between narrowband and broadband services as well as differences among broadband services (ADSL, CATV and FTTH) concerning nominal access speed.

In what follows, we scrutinize the details of the survey. The reasons for choosing their present Internet access service are shown in Figure 2.23: (1) always-on connectivity (55.9 per cent); (2) a flat-rate system (41.0 per cent); (3) low prices (31.7 per cent); (4) transmission speed (25.7 per cent); and such miscellaneous reasons as easy introduction, no-charge campaigns, IP telephony and CATV service.

Figure 2.24 gives the reasons for using the Internet: (1) Web browsing

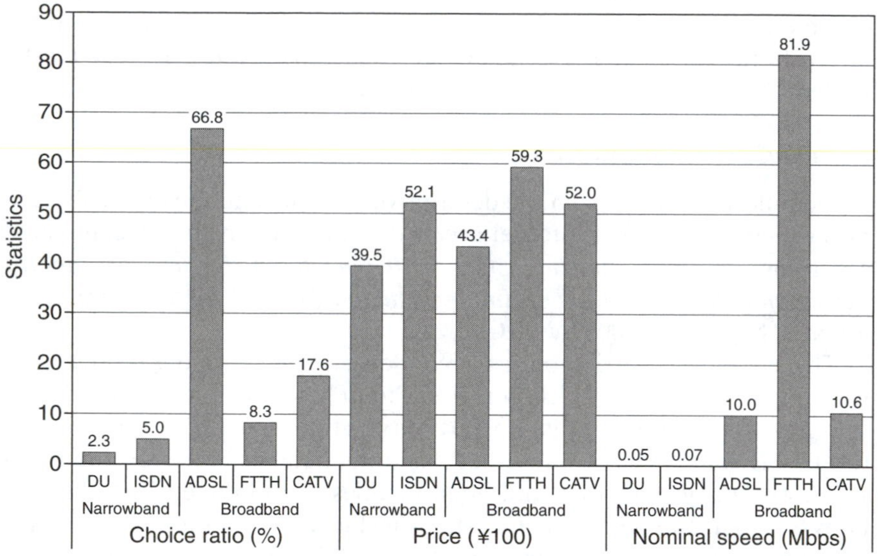

Figure 2.22 Basic statistics concerning broadband markets (source: Ida and Kuroda, 2004, Table 1, p. 24).

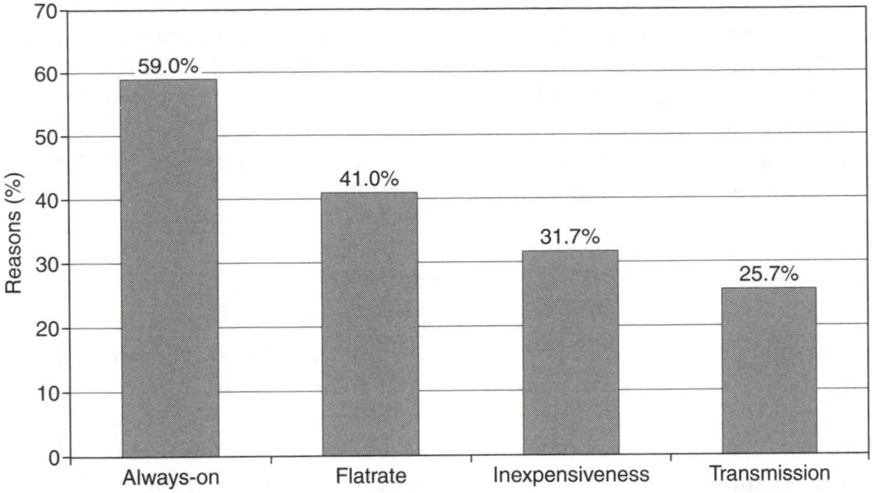

Figure 2.23 Reasons for choosing Internet access service (source: Ida and Kuroda, 2004, p. 9).

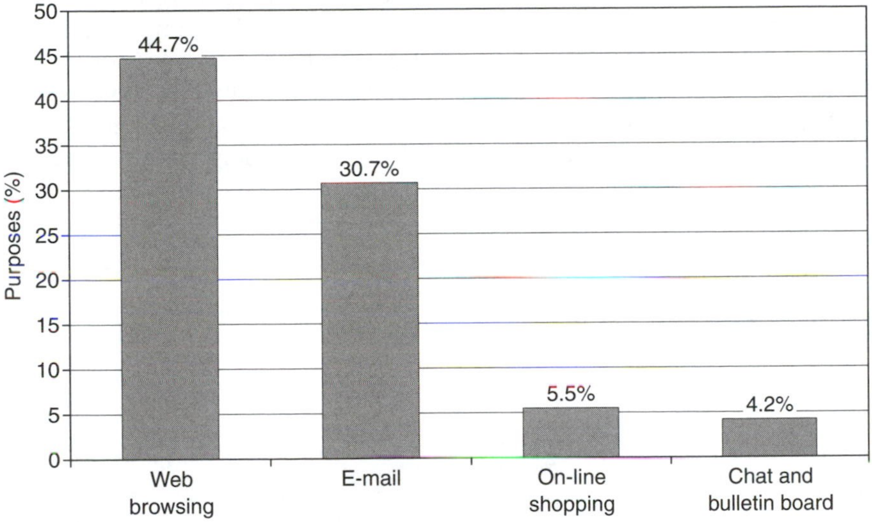

Figure 2.24 Purposes for seeking Internet access (source: Ida and Kuroda, 2004, p. 10).

(44.7 per cent); (2) email (30.7 per cent); (3) online shopping (5.5 per cent); and (4) online chat and bulletin boards (4.2 per cent). Multiple answers were permitted for the above questions.

We also examined the differences between NTT and non-NTT users. On average, NTT users pay about ¥1,000~¥1,500 more per month than non-NTT users: NTT dominates the market despite higher Internet access fees. NTT users still account for 32 per cent of the ADSL and 65 per cent of the FTTH markets. The results of a poll of choice criteria of current Internet access providers (allowing for multiple answers) are summarized in Figure 2.25: (1) low price (44.4 per cent), (2) brand power (23.0 per cent), (3) access speed and functionality (22.7 per cent), (4) stability and reliability (17.9 per cent). Internet service users can be divided into two groups: non-NTT users, who mainly focus on price aspects; and NTT users, who emphasize brand power or reliability.

Nested logit model analysis of broadband services

Next we examine the estimation results of a discrete choice model analysis of broadband service demand. First it will be helpful to refer to Taylor's informative survey.[2] One innovation of demand analysis literature in the 1980s and 1990s is the widespread use of discrete choice models, particularly for analysing access demand based on the assumption that consumer choice is qualitative, with or without access. Perl was one of the first to apply discrete choice models to the analysis of telecommunications access demand, followed by such discrete choice models as logit and probit.[3]

The emergence of a *nested logit model*, which partially alleviated the IIA properties of conditional logit models, was especially important.[4] Ida and Kuroda, however, offered the first comprehensive analysis of broadband service demand including FTTH, the actual broadband service.[5]

We next analysed Internet access demand using a nested logit model based on the work of Ida and Kuroda (2004). Dependent variables are the

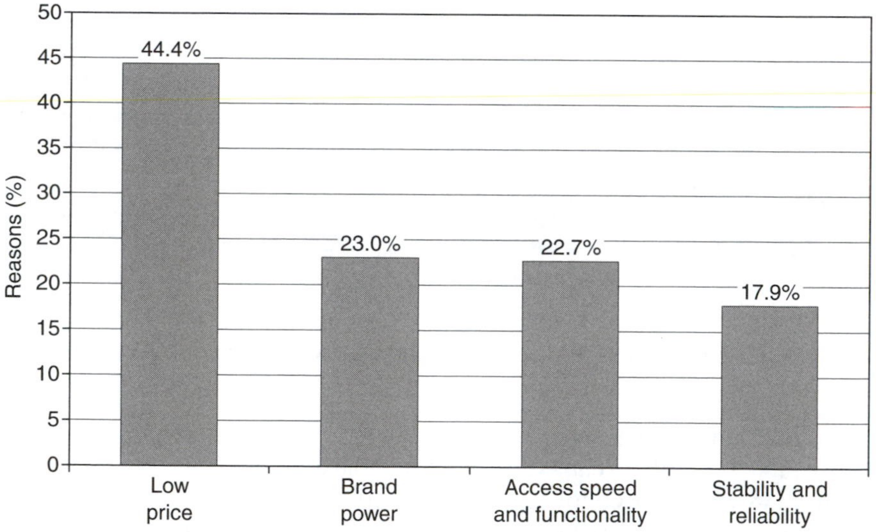

Figure 2.25 Reasons for choosing a particular Internet provider (source: Ida and Kuroda, 2004, p. 11).

four Internet access services: (i) NB (DU/ISDN); (ii) ADSL; (iii) CATV Internet; and (iv) FTTH. Independent variables are: (i) the fixed term of each alternative; (ii) average monthly expenditure (price); (iii) nominal access speed; and (iv) an NTT users dummy variable.

If we adopt a nested logit model, determining the nested choice structure becomes problematic.[6] Thus, we compared the degrees of fitness of the models (that is, the adjusted McFadden R^2) and determined the best model with the highest value. Consequently, we deemed it appropriate to divide the four alternatives into two categories: a narrowband category that includes DU and ISDN, and a broadband category that includes ADSL, CATV, and FTTH.

Estimation results are shown in Table 2.1. Since a McFadden R^2 value of 0.3 generally corresponds to around 0.6 of OLS R^2, we concluded that a McFadden R^2 value of 0.49 represents a high degree of fitness. Fixed terms and price parameters are statistically significant, but not nominal speed parameters. Furthermore, although we included an NTT dummy variable in the model because NTT users pay significantly more than non-NTT users, this statistical significance is very low.

Let us next consider the own-price elasticities of access demand, as summarized in Figure 2.26. The ADSL figure is about 0.3, and thus its service is very inelastic. An increase in ADSL price does not significantly decrease the demand for ADSL. Its market itself is so gigantic that ADSL users are switching from low-speed (1.5 Mbps) to medium-speed (8~12 Mbps) and finally to high-speed band (more than 24 Mbps) within the ADSL market. Below, we scrutinize the ADSL market and divide it into three submarkets. On the other hand, own-price elasticities of access demand are about 1.1 for FTTH and 0.9 for CATV, which are borderline cases. This means that a 1 per cent increase in price induces almost the

Table 2.1 Estimation results of a nested logit model of broadband services

log L(β)	−690.6425		
Log L(0)	−1,367.8963		
McFadden R^2	0.49363		
Parameters	*Coefficients*	*Standard errors*	t-*values*
Fixed term (NB)	−1.66241	0.25424	−6.53879
Fixed term (FTTH)	−1.65073	0.65805	−2.50853
Fixed term (CATV)	−0.74102	0.31161	−2.37803
Price	−0.00021	0.00006	−3.56811
Nominal speed	0.0000043	0.0000041	1.04347
NTT dummy	17.82896	4155010	0.00000
IV (BB)	1.00000	0.35760	2.79645

Source: T. Ida and T. Kuroda, 'Discrete Choice Analysis of Demand for Broadband in Japan', Graduate School of Economics, Kyoto University, COE21 Discussion Paper No. 37, p. 24 (2004).

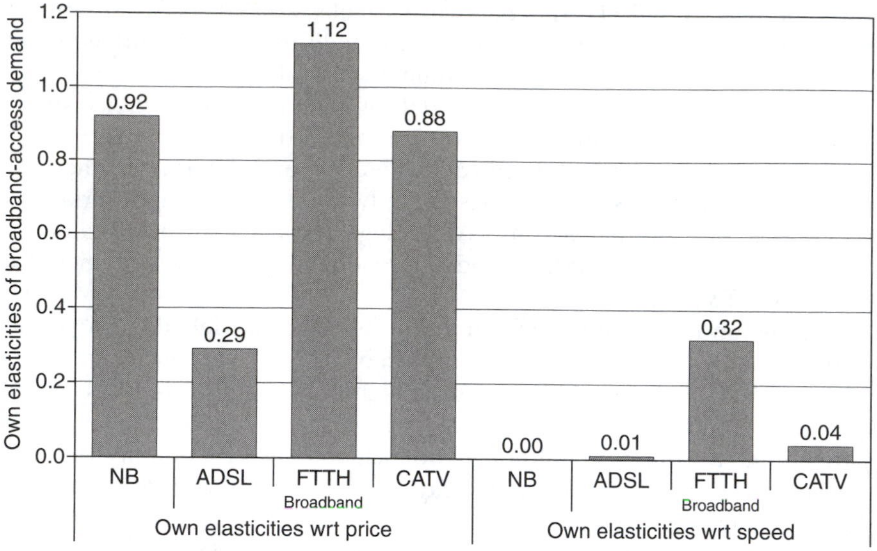

Figure 2.26 Own elasticities of broadband access demand (source: Ida and Kuroda, 2004, Table 4, p. 27).

same decrease in demand. Consequently, ADSL service is less elastic with respect to price than FTTH and CATV, even though the latter two are not highly elastic.

Nested logit model analysis of ADSL services

In the previous section, we showed that the own-price elasticity of ADSL demand is much lower than that for FTTH and CATV. However, since the ADSL market itself is so huge, occupying around 70 per cent of the entire broadband market, it is informative to examine the submarkets of ADSL.

At this point, we divide the ADSL market into three submarkets: low-speed (around 1.5 Mbps), medium-speed (around 8–12 Mbps) and high-speed (more than 24 Mbps). Note that medium-speed ADSL users account for 74 per cent of the ADSL market. However, medium-speed ADSL users are expected to switch to high-speed ADSL and eventually to FTTH. In fact, respondents to a questionnaire concerning Internet access services said that in the near future they want to use the following (multiple answers admitted): (1) FTTH (74.7 per cent); (2) ADSL (35.2 per cent); (3) CATV (29.8 per cent); and (4) fixed wireless access (FWA) (12.6 per cent).

Looking at the own-price elasticities of access demand, shown in Figure 2.27, the medium-speed ADSL figure is about 0.15, which is very inelastic. On the other hand, the figures are 7.3 for low-speed ADSL and 6.7 for high-speed ADSL, which is quite elastic. In conclusion, ADSL users can

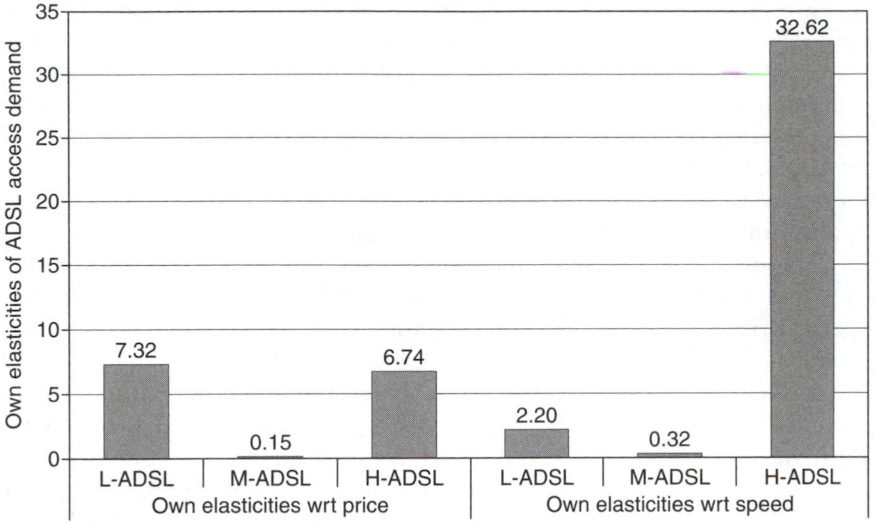

Figure 2.27 Own elasticities of ADSL access demand (source: Ida and Kuroda, 2004, Table 8, p. 31).

be divided into two groups: those who are insensitive to price changes, the medium-speed ADSL users, and those who are very sensitive to price changes, the low- and high-speed ADSL users.

Conjoint analysis of broadband markets

This section investigates the stated preference of broadband markets based on Ida and Sato (2004) (see note 1). We first explain the conjoint analysis used in the analysis. We next discuss the estimation results. Consequently, we found that the actual availability of FTTH has an effect on the stated preferences of consumers and also that stated preference and revealed preference may vary for certain populations.

Conjoint analysis

The key approach adopted here is *conjoint analysis*, or the stated preference method (SPM). It aims to measure consumers' preferences based not on the actual data observed in the market, but instead on answers to a virtual questionnaire, which highlights a remarkable difference from revealed preference methods (RPM). Compared with RPM, one advantage of SPM is that it is virtual by nature and scrutinizes consumers' preferences that are generally ignored or overlooked. For instance, even when we cannot collect actual product data prior to a product's market launch, we can quantitatively evaluate products with SPM.

Because of such merits, SPM has been utilized in market research fields for product development and demand forecasting. Additionally, it is applied to such non-market goods as environment and health, the market prices of which are difficult to establish through market mechanisms. Although little research has been done in the field of telecommunications, a noteworthy exception is Madden and Simpson, who studied residential broadband subscription demand using SPM.[7] Following them in a methodological sense, Ida and Sato's study is the first comprehensive application of SPM to FTTH, the actual broadband service.

Conjoint analysis assumes that goods or services are composed of attributes that can be profiled. For example, in a broadband service context, speed, price, the availability of IP phones, the distribution of TV programmes, and the symmetry between uploading and downloading functions are considered attributes. Particular analytical purposes shape the contours and number of attributes introduced into a profile. If we include too many attributes, respondents will have difficulty answering the questions. On the other hand, if we incorporate too few attributes, the description of data will become inadequate. Accordingly, we need to repeatedly conduct pre-tests and carefully observe consumer recognition patterns before determining adequate profiles. The profiles used in Ida and Sato's study were made from the combination of attributes and levels enumerated in Table 2.2.

Data and descriptive statistics

Next we discuss the data and the descriptive statistics collected by the same survey based on *The Guidelines* and *The Implementation Manual* as explained earlier in the chapter. The survey was carried out as a Web questionnaire because using the Internet was inexpensive and quick, and because the object of this research itself was related to Internet access. Respondents were randomly chosen from people who have access to all

Table 2.2 Profiles of conjoint analysis

Attributes	*Level*					
Price	¥2,500	¥3,000	¥3,500	¥4,000	¥4,500	¥5,000
	¥5,500	¥6,000	¥6,500	¥7,000	¥7,500	
Access speed	1 M	10 M	20 M	30 M	100 M	
IP telephony	Available	Unavailable				
TV programmes	Available	Partially available		Unavailable		
Provider	NTT (East and West)		Non-NTT			
Symmetry	Symmetric	Asymmetric				

Source: T. Ida and M. Sato, 'Conjoint Analysis of Consumer Preferences for Broadband Services in Japan', Graduate School of Economics, Kyoto University, COE21 Discussion Paper No. 32, table 1, p. 18 (2004).

five Internet alternatives: (i) dial-up Internet (DU); (ii) always-on ISDN; (iii) ADSL; (iv) FTTH; and (v) CATV Internet. The total number of samples was 1,013. Ida and Kuroda (2004) analysed RP for broadband based on the data. The present study also utilizes the same survey for people who do not have access to FTTH.

We then surveyed a random sample of two groups to investigate how actual FTTH availability influences SP, or how the SP and RP of identical populations are different. Group A, derived from a population having access to all alternatives, is 105. Group N, derived from the population without access to FTTH, is 104.

If we look at the answers from all samples, as indicated in Figure 2.28, the stated choices of the respondents (with SPM) break down as follows: (1) ADSL (57 per cent); (2) CATV Internet (13 per cent); and (3) FTTH (30 per cent). On the other hand, Ida and Kuroda (2004) report that the actual choices (with RPM) are (1) ADSL (72 per cent); (2) CATV Internet (19 per cent); and (3) FTTH (9 per cent). It is interesting that ADSL is overwhelmingly supported in SPM as well as in RPM, but the ratio of those choosing FTTH in SPM is higher than in RPM.

Next, if we compare the two subsamples of groups A and N, the ratio of choosing ADSL is higher in group N (60.1 per cent) than in group A (52.5 per cent); on the other hand, choosing FTTH is lower in group N (27.6 per cent) than in group A (32.5 per cent). Accordingly, we assume that the actual availability of FTTH would influence consumers' preferences and cause different choice behaviour in this conjoint analysis.

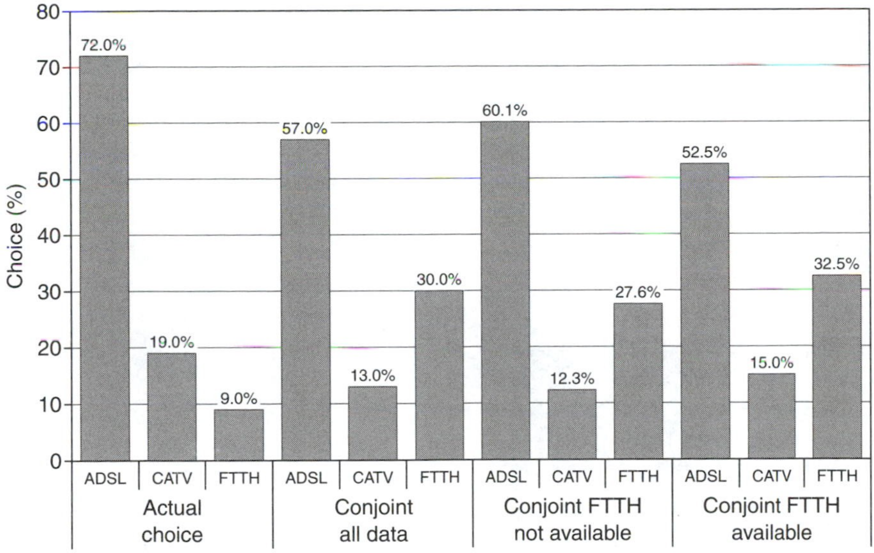

Figure 2.28 Choice results of conjoint analysis (source: Ida and Sato, 2004, Tables 2, 3, 4, pp. 20–22).

Influence of FTTH availability on stated preference

Conjoint analysis studies consumer preferences on the basis of consumers' virtual choices. It is interesting to consider whether respondents' choices are influenced by the actual availability of alternatives. As stated, the respondents were divided into two subsamples: group A with access to FTTH, and group N without, a division merely based on the difference of living environments of respondents; otherwise the questionnaire was identical for the two groups. If this difference in the actual availability of FTTH systematically influences the consumers' SP, the estimated coefficients will be different between the two groups.

Figure 2.29 summarizes the estimation results of groups A and N – that is, the values of WTP. The WTP for 1 Mbps is about ¥30 ($0.27) for group A individuals who have access to FTTH, while it is about ¥70 ($0.63) for group N without. People without access to FTTH have a higher preference for an increase in access speed than those to whom FTTH is available.

FTTH is mostly available in such urban areas as Tokyo, Osaka and Nagoya; areas without access to FTTH are mainly rural or sparsely populated. In urban areas, the competition between firms that provide ADSL and FTTH is fierce; therefore, people living in urban areas can easily switch services or providers. On the other hand, there is little or no competition in rural areas, where a single firm – or at worst no firm – is providing broadband services. Therefore, since it is important for group A to locate better, cheaper broadband services, they are more aware of price and speed; group N are less concerned, because they are primarily concerned with securing access to broadband services.

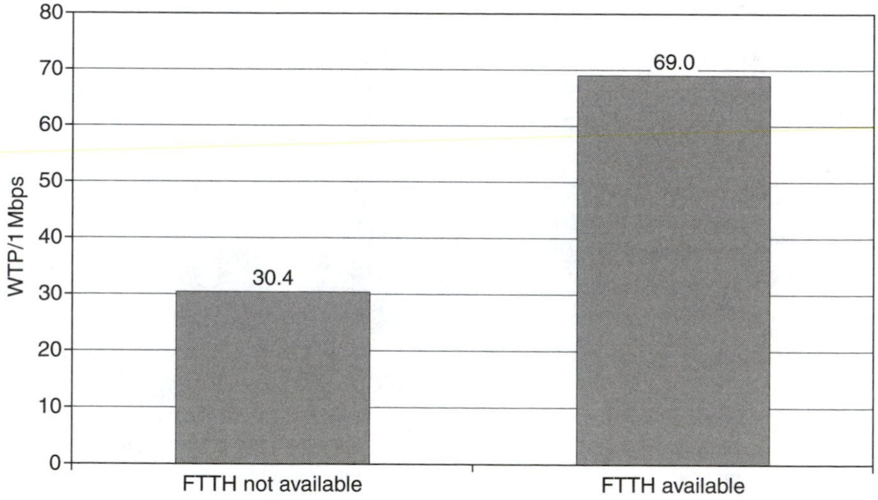

Figure 2.29 FTTH availability and willingness to pay for speed (source: Ida and Sato, 2004, Table 6, p. 24).

Comparison of stated and revealed preferences

We next investigate whether the SP and RP of identical respondents are different. This chapter has so far analysed the SP of broadband services. For a comparative analysis of SP and RP, we use the SPM results of group A derived from the present study and the RPM results of Ida and Kuroda's study.

According to previous research, SP and RP do not always correspond with each other, even though based on random samples from the same population. Carson *et al.* introduced examples that suggest the ratio of an RP value to an SP value varies from 0.005 to 10.269.[8]

As shown in Figure 2.30, SP and RP are clearly divergent. As for the WTP of 1 Mbps, the figure of SP is about ¥30 ($0.27), and RP is about ¥20 ($0.18). Why is the WTP of SP higher than RP? Azevedo *et al.* discussed a similar example in which the WTP of SP is two and a half times higher than the WTP of RP.[9] They argue that respondents tend to consider income constraints 'softer' in SP than in RP. Besides, as Louviere *et al.* state, since SPM primarily considers innovative or qualitative changes of goods or services, SP is thought to indicate not a temporary but a long-term preference.[10] In this light, it is not surprising that in rapidly developing broadband services the WTP of SP is one and a half times higher than RP.

Further discussions: future problems

The broadband market in Japan took off more quickly than in any other country, to some extent by chance, yet it was also inevitable. In this final

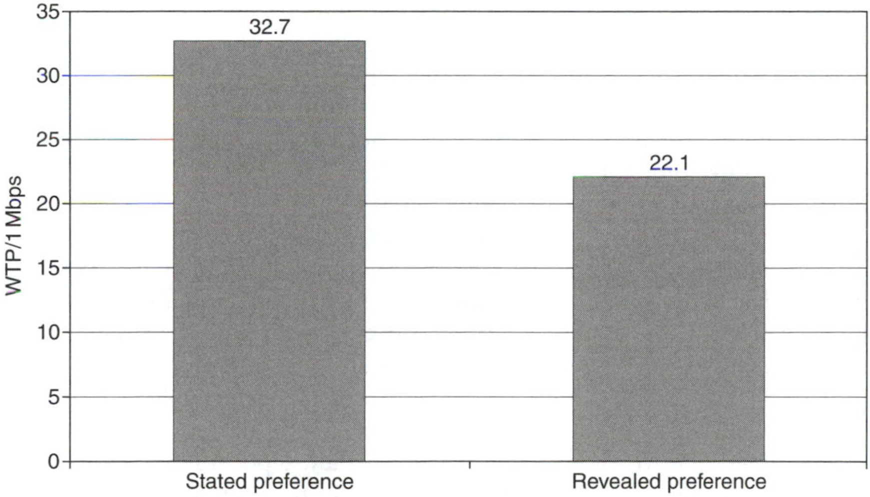

Figure 2.30 Discrepancies between stated preference and revealed preference for broadband services (source: Ida and Sato, 2004, Table 6, p. 24).

section, instead of summing up the previous discussions, we consider four future problems faced by Japan's broadband industry.

The first is a problem confronting the Ministry of Internal Affairs and Communications. When it was called the Ministry of Posts and Telecommunications, MIC was regarded as second-rate; now, twenty-odd years after telecommunications liberalization, it has become a leader in information/communications policies. There is, however, a matter of concern. Such industrial policies as universal service support and the convergence between communications and broadcasting are becoming increasingly important issues that will not be resolved only by competition policies. It seems difficult even for MIC, a first-rate ministry today, to bear the heavy responsibilities of both competition and industrial policies in the telecommunications industry.

The second is a problem facing NTT. Japan's broadband market has succeeded because, on a national scale, NTT has ubiquitously provided telecommunications services, including broadband. Sooner or later, after large numbers of customers, mainly business users, replace POTS with IP phones, the telephone revenues of NTT East and NTT West will decline sharply, and their publicly switched telephone networks will not be maintained. Since fierce price competition has already started in the mobile phone market, NTT DoCoMo cannot afford to subsidize NTT East and NTT West any longer. One possible outcome suggests that NTT East and West will not hold the telephone networks for ever and will replace telephone networks with well-planned, high-speed IP broadband networks.

The third is a problem of new entrants. Japan's success in broadband Internet diffusion owes a great deal to the new businesses of various entrants. However, their profitability is extremely low. If such 'upstart entrants' are beaten and chased from the market, the effects of market mechanisms will be very limited. In particular, we should closely follow the future fortunes of Softbank, which is different from NTT and even the other entrants because it does not have a strong business financial supporter. Even though opinions differ over Softbank's business strategy, the explosive development of Japan's broadband market would have been greatly delayed without its surprising entry.

The fourth point concerns the demand side. Even if information can interactively and physically flow at 100 Mbps on optical fibres to homes, possessing such capability is meaningless in the absence of rich, valuable content to be exchanged. Unfortunately, current broadband services offer only cheap substitutes for existing services without creating new values for society. In the future, all of us in the industry must use broadband technology to create higher social values through such services as telemedicine and tele-education.

We would like to comment on the final point. Optical fibre services are not final consumer goods but intermediate inputs. Therefore, they do not produce great effective demands in society themselves. At this point, IP telephony and high-definition (HD) digital broadcasting have been proposed as candidates of killer applications or killer contents for FTTH services, but it seems that they are too indecisive to become the driving force behind broadband services. There is a strong possibility that telemedicine and tele-education – rather than commercial applications – will create greater social value in fields where the latest information and communications technologies have not yet been fully utilized. We will introduce a few examples below.

In Tottori prefecture, they have constructed an information communication network for educational use and started such programmes as information sharing between schools and homes, learning through the Internet, manufacturing educational content, long-distance learning, using groupware for students, and so on.

In Niigata prefecture, they have established a support system for telemedicine and started such programmes as radiation diagnostic imaging, disease case studies via videoconferencing, and information provision for physicians living in remote areas.

Tottori and Niigata are typical rural areas far from Tokyo with a high proportion of senior citizens. It is interesting that innovations have started in areas where the construction of broadband networks lags behind that in urban areas. This fact indicates that the current scale of the broadband market does not always coincide with the potential size of the social values created by broadband services. In this respect, broadband services can become more than expensive playthings for urban youth, including highways to schools and hospitals for rural children and senior citizens. It may seem paradoxical, but if the broadband networks were constructed ubiquitously, the large regional disparity among education and medical services could be solved.

The regulatory reform of Japan's telecommunications used to import experiences from overseas. In broadband services, however, Japan has to find its own path. Imitation is easy, but innovation is difficult. From now on, Japan's broadband market faces a crucial juncture.

Notes

Thanks are due to the following people. Professors Dr. Ruth Taplin and Masako Wakui, the editors of this book, gave me careful advice. Hideyuki Ohashi, Kazuhiro Mita, Syohei Shoji and Toshio Kawamoto of the Ministry of Internal Affairs and Communications worked out *The Effective Competition Review of Japan's Telecommunications* (June 2004), to which this chapter is much indebted. Toshihumi Kuroda, Masayuki Sato and Masahito Anbashi of the Graduate School of Economics, Kyoto University, assisted our joint research with great energy. Furthermore, part of this work first appeared in *Yasashii Keizaigaku* (Easy Economics): 'Information–Telecommunication and Competition Policy' (July 2004) in *Nihon Keizai Shimbun*, the Japanese economic newspaper.

1 See T. Ida, and M. Sato, 'Conjoint Analysis of Consumer Preferences for Broadband Services in Japan', Graduate School of Economics, Kyoto University, COE21 Discussion Paper No. 32 (2004); T. Ida and T. Kuroda, 'Discrete Choice Analysis of Demand for Broadband in Japan', Graduate School of Economics, Kyoto University, COE21 Discussion Paper No. 37.

2 L. D. Taylor, 'Customer Demand Analysis.', in M. E. Cave, K. Majumdar and I. Vogelsang (eds) *Handbook of Telecommunications Economics*, vol. 1, (Amsterdam: North-Holland, 2002).

3 See Perl's articles: L. J. Perl, 'Economic and Social Determinations of Residential Demand for Basic Telephone Service.' National Economic Research Associates (New York: White Plains, 1978); L. J. Perl, 'Residential Demand for Telephone Service 1983', (White Plains, NY: National Economic Research Associates, 1983). Examples of discrete choice models, including nested logit model, include the following: H. Ben-Akiva. and S. Gershenfeld, *Analysis of Business Establishment Choice of Telephone System* (Cambridge, MA: Cambridge Systematics, 1989); D. J. Kridel, 'A Customer Surplus Approach to Predicting Extended Area Service (EAS) Development and Stimulation Rates', *Information Economics and Policy* 3: 379–390 (1988); L. D. Taylor and D. J. Kridel, 'Residential Demand for Access to the Telephone Network', in A. de Frotenay, M. H. Shugard and D. S. Sibley (eds) *Telecommunications Demand Modeling* (Amsterdam: North-Holland, 1990); D. J. Kridel and L. D. Taylor, 'The Demand for Commodity Packages: The Case of Telephone Custom Calling Features', *Review of Economics and Statistics* 75: 362–367; J. Bodnar, P. Dilworth and S. Iacono, 'Cross-section Analysis of Residential Telephone Subscription in Canada', *Information Economics and Policy* 3: 311–331 (1988); D. L. Solvason, 'Cross-sectional Analysis of Residential Telephone Subscription in Canada using 1994 Data', *Information Economics and Policy* 9: 241–264 (1997).

4 See K. E. Train, D. L. McFadden and M. Ben-Akibva, 'The Demand for Local Telephone Service: A Fully Discrete Model of Residential Calling Patterns and Service Choices', *RAND Journal of Economics* 18: 109–123 (1987); K. E. Train, M. Ben-Akibva and T. Atherton, 'Consumption Patterns and Self-Selecting Tariffs', *Review of Economics and Statistics* 71: 62–73 (1989).

5 Although analyses of broadband demands are limited, the following are noteworthy pioneers in the field: G. Madden, S. G. Savage and G. Coble-Neal, 'Subscriber Churn in the Australian ISP Market', *Information Economics and Policy* 11: 195–208 (1999); G. Madden and M. Simpson, 'Residential Broadband Subscription Demand: An Econometric Analysis of Australian Choice Experiment Data', *Applied Economics* 29: 1073–1078 (1997); J. Eisner and T. Waldon, 'The Demand for Bandwidth: Second Telephone Lines and One-line Services', *Information Economics and Policy* 13: 301–309 (2001); D. J. Kridel, P. N. Rappoport and L. D. Taylor, 'An Econometric Model of the Demand for Access to the Internet by Cable Modem', in D. G. Loomis and L. D. Taylor (eds) *Forecasting the Internet: Understanding the Explosive Growth of Data Communications* (Norwell, MA: Kluwer Academic, 2001); K. T. Duffy-Deno, 'Business Demand for Broadband Access Capacity', *Journal of Regulatory Economics* 24: 359–372 (2003).

6 See W. H. Greene, *Econometric Analysis*, 5th edn (New York: Prentice Hall, 2003).

7 G. Madden and M. Simpson, 'Residential Broadband Subscription Demand: An Econometric Analysis of Australian Choice Experiment Data', *Applied Economics* 29: 1073–1078 (1997).

8 See R. Carson, N. Flores, K. Martin and J. Wright, 'Contingent Valuation and Revealed Preference Methodologies: Comparing the Estimates for Quasi-public Goods', *Land Economics* 72: 80–99 (1996).

9 See C. D. Azevedo, J. A. Herriges and C. L. Kling, 'Combining Revealed and Stated Preferences: Consistency Tests and Their Interpretations', *American Journal of Agricultural Economics* 85: 525–537 (2003).

10 See J. J. Louviere, D. A. Hensher and J. D. Swait, *Stated Choice Methods: Analysis and Application* (Cambridge: Cambridge University Press, 2001).

3 The mobile phone industry

A microcosm of deregulation, globalization and technological change in the Japanese economy

Jeffrey L. Funk

Introduction

The interaction between deregulation, globalization and technological change is a major issue in the economics and management literatures.[1] For example, the forces of deregulation and globalization accelerated the international diffusion of the Internet. Deregulation began in the telecommunications sector in the early 1980s and it and the globalization of finance, corporate governance and trade had become powerful forces by the 1990s. Nevertheless, the extent to which the Internet pierced borders depended on the national specificities of law and regulation, business networks, competition, and technological legacies.[2]

Multinational corporations have played key roles in the globalization of the Internet and other industries. US investment banks, consulting companies, venture capitalists, content providers and technology providers have played this role in the Internet,[3] GM, Ford and Toyota have done this in the automobile industry, and Nokia, Ericsson and Motorola have played this role in the mobile phone industry.[4]

The interaction between globalization and deregulation has also been a major issue in the literature on the Japanese economy and other aspects of Japanese politics. The literature on Japanese economics and politics often uses the term *gaiatsu*, which literally means foreign pressure, to describe the impact of foreign political pressure on decisions by the Japanese government. Japanese politicians and individual ministries also use *gaiatsu* to strengthen their own domestic agendas. Many observers have argued that deregulation and globalization would not have proceeded, and will not proceed, in Japan without *gaiatsu*.[5]

This chapter discusses the interaction between deregulation, globalization and technological change in the Japanese mobile phone industry. A key aspect of *gaiatsu* in the mobile phone industry has been the gradual realization by Japanese firms and government agencies that globalizing and working with foreign firms is a necessary tool for competition in the Japanese domestic market. Without the early emergence of this realization, it is likely that globalization and deregulation would have proceeded much more slowly than they actually did in the Japanese market. Furthermore, without globalization and deregulation, it is unlikely that Japanese firms such as NTT DoCoMo and KDDI would have been able to transform themselves into global players.

This chapter is largely based on my previous books and articles on the mobile phone industry. These include a historical analysis of competition in the global market,[6] the Japanese mobile Internet,[7] and more specific analyses of global competition between mobile phone standards, phones[8] and mobile Internet services.[9] These publications provide information on the original sources, which include Japanese (e.g. the *Nikkei Economic Journal*) and English published sources, and interviews with more than 200 managers from firms in Japan, Europe and the United States between 1996 and 2004.

This chapter first discusses models of technological change and the application of these models to the mobile phone industry, which includes a historical overview of the Japanese mobile phone industry. This is followed by separate discussions of each generation of technology, including analogue, digital, PHS, 3G and the mobile Internet.

Technological change and mobile phones

Technological discontinuities lead to a period of ferment in which alternative product forms compete for dominance.[10] This competition between alternative product forms may occur in committees or in the marketplace where one result is the emergence of one or at the most a limited number of standards or dominant designs. The choice of the standard or dominant design, which can include domestic and foreign ones, has a strong effect on the competition between firms in local and global markets.[11]

The mobile phone industry has experienced three major technological discontinuities, and these discontinuities have led to the adoption of different standards; the most important is the air interface standard, which defines how signals are transmitted between phones and base stations. Roughly speaking, dividing an area into cells, reusing the frequency spectrum in each cell, and using analog signals to make calls was the first technological discontinuity. The use of digital signals and the combination of digital and broadband are usually defined as the second and third technological discontinuities. In each generation of phones, there have also been rapid reductions in the size, weight and cost of the phones through improvements in discrete components and integrated circuits.[12]

Table 3.1 summarizes the evolution of the major mobile phone service providers in Japan, their names, and their operating regions. NTT (which spun off NTT DoCoMo in 1992) started services in 1979 and by the mid-1980s it was operating a national service. DDI Cellular and IDO started services in 1988 and 1989 respectively in different regions of Japan, thus resulting in two service providers per region. Digital Phone and Tsuka

Table 3.1 Evolution of service provider names and their operating regions

Year	*Names of service providers at time of formation*					
	NTT	*DDI Cellular*	*IDO*	*Digital Phone*	*Digital Tsuka*	*Tsuka Cellular*
1979	Tokyo, Osaka					
1985	Nationwide					
1988, 1989		Regions outside of Tokyo and Nagoya	Tokyo, Nagoya			
1992	Name changed to NTT DoCoMo					
1994				Tokyo, Nagoya, Osaka	Regions outside of Tokyo, Osaka, Nagoya	Tokyo, Nagoya, Osaka
1999				Merger and name changed to J-Phone		Acquired by DDI
2000		Merger between DDI, IDO, and KDD to form KDDI				
2001		Adoption of au brand name in mobile business		Becomes part of the Vodafone group		
2003				Name changed to Vodafone		

Source: Home pages of firms and author's analysis.

Note
NTT: Nihon Telegraph and Telephone.

Cellular started services in 1994 in Japan's three major metropolitan areas (Tokyo, Osaka and Nagoya). Although they are not shown in Table 3.1, the major investors in these service providers also started digital services that are based on a 'low mobility' technology called Japan's Personal Handyphone System (PHS) in 1995 (see Table 3.2).

To return to the major mobile phone service providers in Table 3.1, Digital Tsuka (a combination of investors in Digital Phone and Tsuka Cellular) began providing services in the regions outside of Tokyo, Osaka and Nagoya and providing roaming services for Digital Phone and Tsuka Cellular in 1997. After changing its name from Digital Phone, J-Phone acquired Digital Tsuka in 1998 and it was acquired by Vodafone in 2001. DDI Cellular, IDO and KDD, which is a provider of international wireline services, merged in 2000 to become KDDI; KDDI also acquired Tsuka Cellular in 2000. Since KDDI and Tsuka Cellular use different air interface standards, there is very little integration of services. As of late 2004, there are four major service providers and three PHS service providers.

Table 3.2 summarizes the air interface standards that Japanese service providers adopted in the three generations of mobile phone services. NTT and IDO used NTT's analogue system while DDI Cellular adopted TACS,

Table 3.2 Service start dates by operator and technology

Technology	*Operator*	*Air interface standard*	*Service start dates (some regions had later start dates)*
1G: Analogue	NTT DoCoMo	NTT	1979
	IDO	NTT	1988
	DDI Cellular	TACS	1989
	IDO	TACS	1991
2G: Digital	NTT DoCoMo	PDC	1993
	J-Phone, Tsuka Cellular, DDI Cellular	PDC	1994
	IDO	PDC	1995
	NTT Personal, DDI Pocket, Astel	PHS	1995
	Digital Tsuka	PDC	1997
	DDI Cellular	cdmaOne	1998
	IDO	cdmaOne	1999
3G	KDDI	cdma20001x	2002
	NTT DoCoMo	W-CDMA	2002
	Vodafone	W-CDMA	2003

Source: Home pages of firms and author's analysis.

Note
TACS: Total Access Communication System; PDC: Personal Digital Cellular; PHS: Personal Handyphone System; cdma: code division multiple access; W-CDMA: wide-band code division multiple access.

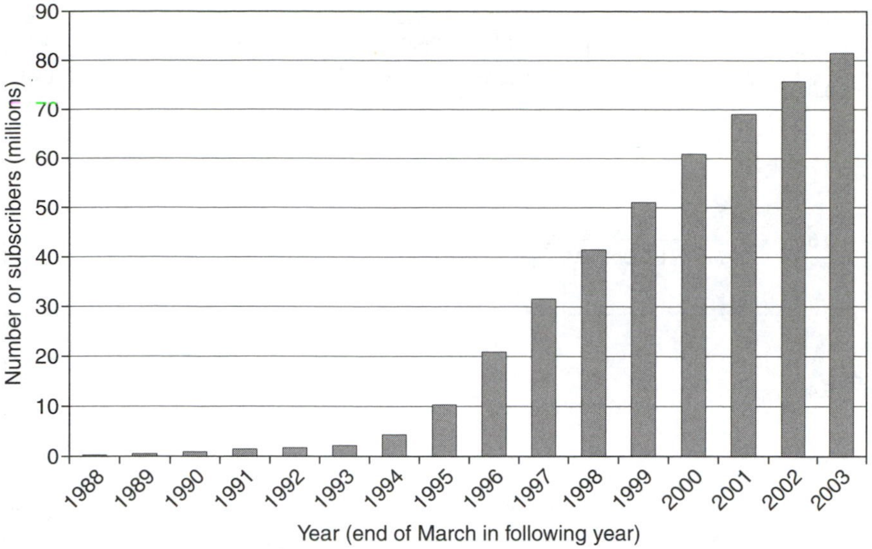

Figure 3.1 Growth in the number of mobile phone subscribers (source: Japan's Telecommunication Carriers Association and author's analysis).

which is a derivative of the US AMPS and the British TACS. NTT DoCoMo began digital services in 1993 based on PDC (Personal Digital Cellular), as did Vodafone, Tsuka Cellular and DDI Cellular in 1994, and Digital Tsuka in 1997. DDI Cellular started services based on cdmaOne in 1998, followed by IDO in 1999.[13] After they merged in 2000 to form KDDI, they renamed the mobile phone service 'au' and started cdma20001x services in April 2002. NTT DoCoMo started FOMA services in January 2002 and Vodafone did so in 2003.

Figures 3.1 and 3.2 summarize the growth in subscribers (including PHS) and the evolution in shares (cumulative data) for the non-PHS service providers. The entry of new service providers stimulated the growth in subscribers in 1994 and, to a lesser extent, in 1989; the former is hard to discern in Figure 3.1, owing to the much lower number of subscribers in 1989 as compared with the present. As shown in Figure 3.2, the entry of new service providers has also had an impact on the share of subscribers.

Analogue: the first wave of *gaiatsu*

NTT started the first mobile phone services in the world in 1979. NTT used a proprietary system that it and a small number of Japanese suppliers had developed. In spite of Japan's two-year lead in these services, other factors prevented the adoption of the standard by other countries:

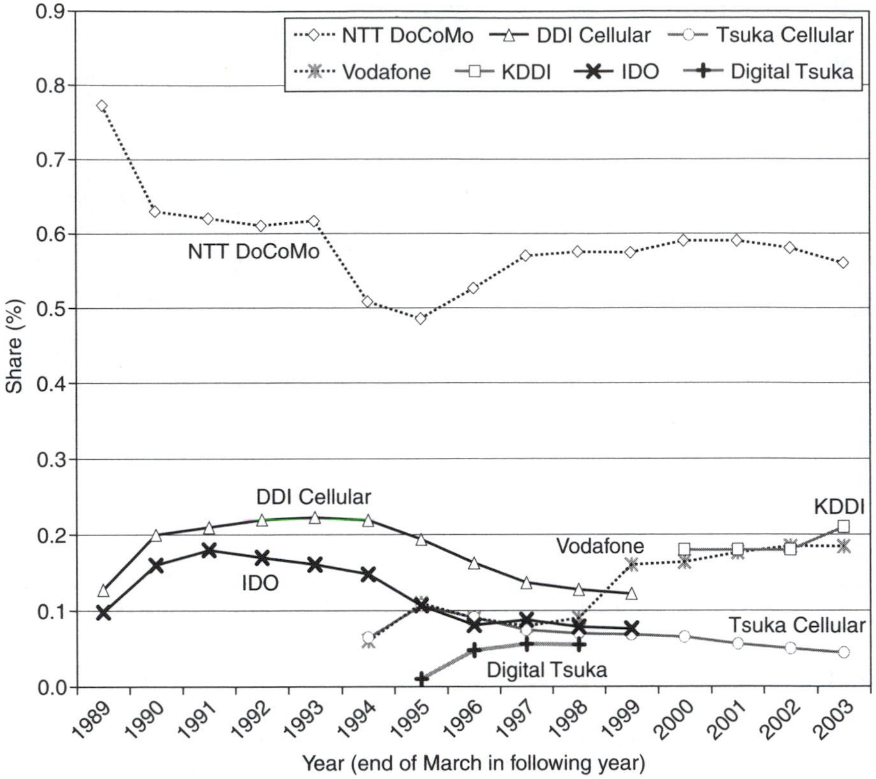

Figure 3.2 Different service providers' shares of mobile phone subscribers (source: Japan's Telecommunication Carriers Association and author's analysis).

these include the closed nature of the standard, the small number of participating manufacturers, which were all Japanese, and the lack of growth in the number of subscribers.[14]

NTT's monopoly in services, the rental as opposed to sale of phones, and the high price of the services led to very slow growth in the number of subscribers. Neither the Ministry of Posts and Telecommunications (MPT) nor NTT believed there was a large market for mobile communication services in Japan and thus NTT did not install many base stations, and both the MPT and NTT agreed to set user fees at high levels. The MPT wanted to set fees high in order to minimize the number of complaints and prevent people from subscribing who could not afford the service, and it rented the phones because it thought the purchase of phones would confuse users.[15] The MPT was apparently influenced by the Ministry of Transportation, whose offices were adjacent to the MPT at the time. The MPT controlled the renting of phones and the awarding of spectrums, just as the Ministry of Transportation controlled vehicle licences. In spite of having less than one-fifth the population of Japan, there were more mobile phone subscribers in Scandinavia than in Japan by the end of 1983.[16]

Gaiatsu, in combination with domestic interest in deregulation, ended NTT's monopoly on mobile phone services by allowing the entry of two new service providers, one of which used foreign technology. Pressure in the mid-1980s from the US government is one reason why Japan's MPT agreed to give one licence to a firm that promised to implement a foreign standard and buy from a US company.[17] IDO, whose major investor was Toyota, adopted the NTT standard and started services in Nagoya and Tokyo in late 1988. Toyota believed that it was safer to adopt a Japanese standard than a foreign one even if the standard was provided by its major competitor (Shigetaka, 1995; see note 17). DDI Cellular, whose major investor was Kyocera, adopted TACS, which is a derivative of the US AMPS standard, purchased the equipment from Motorola, and started services in the other regions of Japan in early 1989.

DDI Cellular's use of an open foreign standard enabled it to obtain 48 per cent of the subscribers in its regions (versus 52 per cent for NTT DoCoMo) as compared to 31 per cent for IDO (versus 69 per cent for NTT DoCoMo) in its regions of operation between the date of their service starts and March 1994. In particular, the large installed base of the AMPS and TACS standard in the United States and elsewhere enabled DDI Cellular to offer handsets that were smaller and cheaper than NTT DoCoMo's handsets. On the other hand, IDO paid NTT a licensing fee for the use of the NTT standard, it did not receive phones until six months after NTT DoCoMo had received the phones, and few suppliers developed handsets based on NTT DoCoMo's analogue standard, because of the small market and high licensing fees.[18]

Motorola, the US government and DDI Cellular continued to argue that the lack of a nationwide service represented a disadvantage to DDI Cellular and thus IDO should also adopt the TACS standard in order to provide roaming services for DDI Cellular. IDO's main investor, Toyota, did not strongly oppose these arguments, partly since it did not want to anger the government of its second largest market. As a result, IDO adopted and gradually implemented TACS equipment in the early 1990s as it simultaneously implemented digital base stations.[19]

Digital: the second wave of *gaiatsu*

Gaiatsu also played a role in the choice of service providers, standards and suppliers for digital services, which were started in the early 1990s in Japan, Europe and the United States. Partly as a result of *Gaiatsu*, the Japanese government awarded two new licences and allowed foreign firms to invest in the new entrants. The largest investment was made by Vodafone, but more than ten other foreign companies made investments in Digital Phone, Tsuka Cellular and, later, in the PHS service providers.

Vodafone's investment in Digital Phone eventually led to its acquisition of this firm (then called J-Phone) in 1999.

In terms of standards, the US government pressure probably played a role in NTT's design change from frequency division (FDMA) to time division multiple access (TDMA) in the late 1980s. Both US and European firms had been focusing on the latter since the mid-1980s. Although NTT claims that the decision to adopt TDMA and also Motorola's coding and decoding technology in 1990 was based solely on technical reasons, it is likely that political factors also played a role, perhaps the largest role.[20]

In addition to the successful use of foreign technology by DDI Cellular in analogue cellular, both political and technical factors probably also played roles in the choice of suppliers for NTT DoCoMo's digital services. As part of the US–Japanese Telecommunications Agreement in 1985, the NTT Procurement Agreement required NTT to purchase 20 per cent of its equipment from foreign suppliers. The decision to change to technology in which foreign firms had greater capabilities than domestic firms and NTT's falling share in the analogue market probably also encouraged NTT to work with foreign suppliers. Ericsson and to a lesser extent Motorola became suppliers of infrastructure, while Nokia became a supplier of handsets.[21]

Partly through pressure from the US government, the Japanese government also required service providers to sell rather than rent phones and required NTT to publish a set of specifications for the air interface and freely license this technology to the other service providers. The much faster subscriber growth in countries like the United States, Great Britain and the Scandinavian countries, which sold phones and adopted open standards, made it hard for the Japanese government to resist these arguments. As part of publishing the specifications for the air interface, NTT was required to obtain approval for the standards from the Japanese Association for Radio Industry Businesses (ARIB).[22]

NTT DoCoMo's revival

In spite of the government's requirements to publish the specifications for the digital standard, however, NTT DoCoMo's used its superior financial resources to control the digital standard and it used this control to obtain smaller and lighter handsets from phone manufacturers than its competitors during the early years of digital services. NTT DoCoMo was the only service provider to develop proposals and test the equipment with a select set of manufacturers. Thus, its competitors have had little choice but to rubber-stamp NTT DoCoMo's proposals, many times just as NTT DoCoMo was implementing them. This standard-setting process is very similar to the ones used in Europe in both wireline and most analogue mobile phone services. The success of the Scandinavian NMT standard-setting process persuaded West European firms to change from this rather closed to a more open standard-setting process for GSM.[23]

NTT DoCoMo used its control of the PDC standard and its close relationships with its handset suppliers to obtain better handsets than its competitors. It provided its four suppliers (Matsushita, NEC, Fujitsu, Mitsubishi) with preferential information about the PDC standard and its planned services in return for preferential access to phones. These suppliers used this preferential information to make better design decisions and to receive preferential treatment from part suppliers. The phone suppliers agreed not to deliver the handsets to other service providers until six months after DoCoMo received them.[24]

NTT DoCoMo's advantages in handsets became very apparent to users upon the release of the first sub-100-gram phone in late 1996, which coincided with a reduction in activation commissions (phone subsidies) by the other service providers, owing to their relatively high cancellation rates. The combination of the sub-100-gram phone and reduced activation commissions caused DoCoMo's share of all new subscribers to rise dramatically in late 1996 (see Figure 3.3). On a monthly basis, DoCoMo's share of

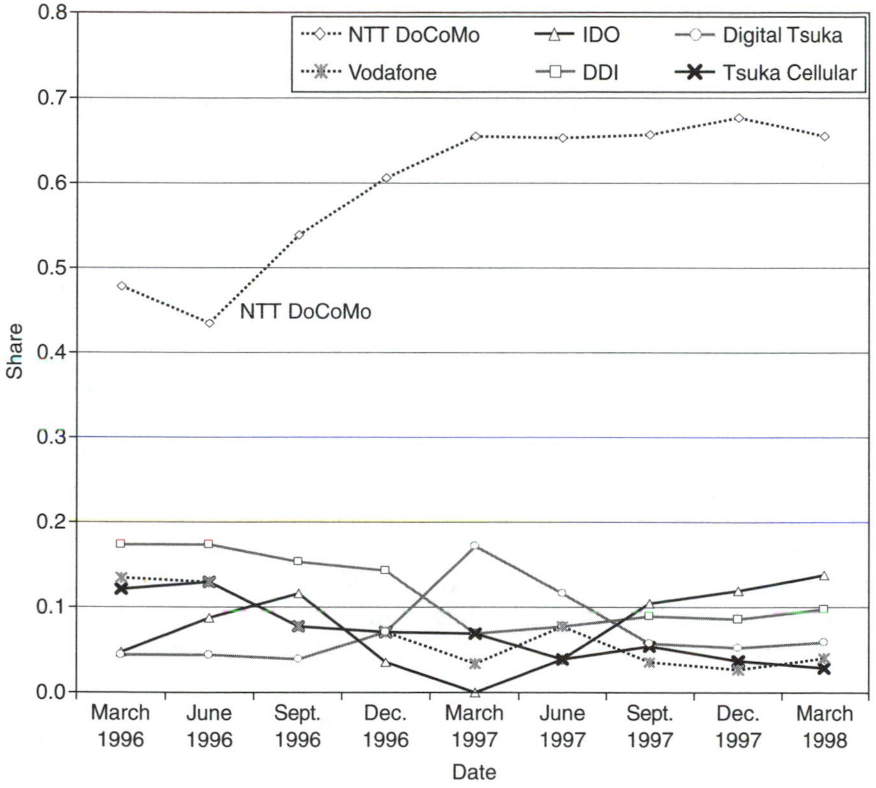

Figure 3.3 Different service providers' shares of net additions in the previous three months, 1996–1998 (source: Japan's Telecommunication Carriers Association and author's analysis).

new subscribers rose from 48 per cent in August to over 60 per cent in October and it stayed over 60 per cent throughout 1997, including a high of 71 per cent in January 1997. Furthermore, NTT DoCoMo's phone suppliers also experienced rising shares, owing to the success of NTT DoCoMo and the demand for their phones from the other service providers.[25]

Importing another foreign standard

The falling shares of DDI Cellular and IDO caused them to begin services based on a second foreign technology called code division multiple access (cdma) in 1998 and 1999 respectively. DDI Cellular had succeeded with a foreign analogue technology and it wanted to return to a situation where it used a technology superior to that used by NTT DoCoMo. Both DDI Cellular and IDO's success with TACS convinced IDO that it should also implement cdmaOne services. Both service providers had been slow to realize that Japanese consumers perceived digital as superior to analogue services and that NTT DoCoMo's control of PDC prevented them from effectively competing in the market.[26]

DDI Cellular started cdmaOne services in Kansai, Kyushu and Okinawa in July 1998 and achieved a nationwide service when IDO started services in Tokai and Tokyo in April 1999. Interestingly, the same month that Kansai Cellular started its cdmaOne services, Kyocera released the smallest PDC phone (69 grams) on the Japanese market, thus eliminating one of the reasons for changing to cdmaOne.[27]

It has taken DDI Cellular and IDO many years to obtain a higher share of new subscribers via the advantages of cdmaOne services. Although the superior voice quality of cdmaOne had led many people to believe that DDI Cellular and IDO would quickly take share away from NTT DoCoMo, this did not happen until KDDI introduced cdma2001x and *Chaku Uta* services in 2002. Initial problems included heavy and expensive phones with shorter battery times than the PDC phones and less than expected benefits from the superior voice quality. While the voice quality of cdmaOne was superior to that of PDC in fixed-line to mobile phone calls, it was similar to that for mobile calls made between PDC phones or between PDC and cdmaOne calls. And by the time DDI Cellular and IDO had introduced cdmaOne, the mobile to mobile calls had already become more frequent than the fixed-line to mobile calls and, because of the wide use of PDC by the other service providers, most mobile to mobile calls by KDDI users were PDC to cdmaOne calls (or vice versa).

Interestingly, Vodafone (then called J-Phone) outperformed both DDI Cellular and IDO in 1999 through the success of its J-Sky messaging service. As shown in Figure 3.4, J-Phone had a higher share of new subscribers than both DDI Cellular and IDO between early 1998 and early 2000. Furthermore, even if the shares for DDI Cellular and IDO are

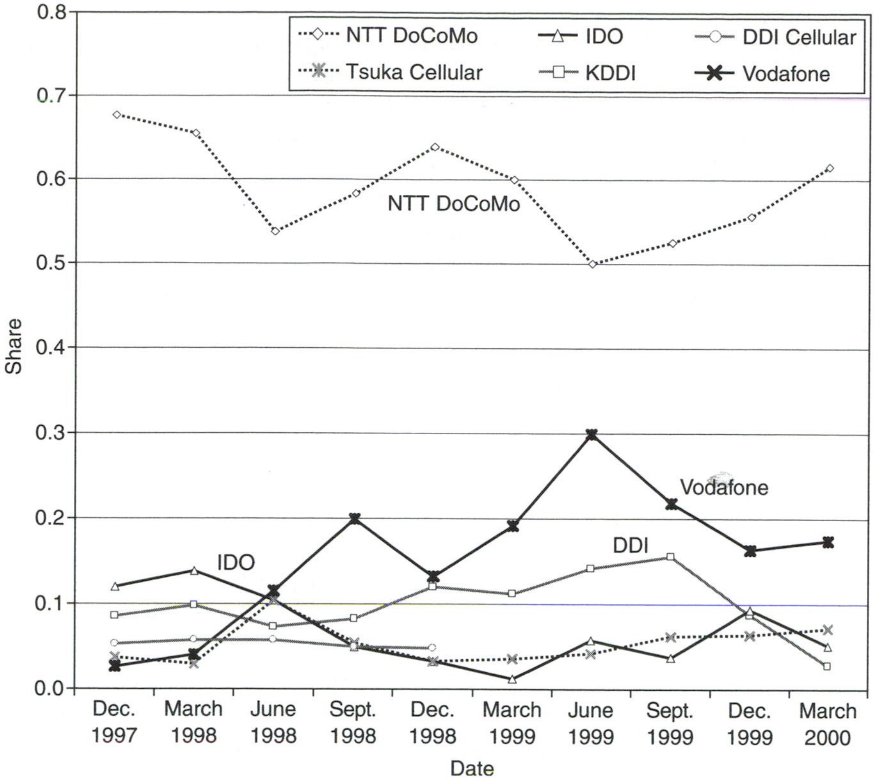

Figure 3.4 Different service providers' shares of net additions in the previous three months, 1997–2000 (source: Japan's Telecommunication Carriers Association and author's analysis).

combined, J-Phone has had a higher share of new subscribers since the former started cdmaOne services in April 1999.

PHS: a partial success in deregulation and a failure in globalization

While NTT DoCoMo was developing and implementing its PDC standard, the MPT and many Japanese manufacturers were also developing a low-mobility digital system called PHS (Personal Handyphone System). PHS is characterized as a low-mobility digital system because calls were initially very difficult to handle in fast-moving vehicles. Because of the very low penetration rate in the Japanese market as compared to many other countries in the early 1990s, many Japanese officials believed that the Japanese market would not experience growth unless a completely new kind of system were introduced. Thus, the Japanese MPT would only approve a system that was one-quarter as expensive as cellular technology, and this directive required Japanese firms to take full advantage of the low cost potential of the small cell, small base station approach.[28]

These technical differences have enabled PHS to become moderately successful in developing countries with high population densities. There were almost 60 million subscribers outside of Japan as of mid-2004,[29] of whom 54 million were in China (EMC, 2004).[30] China's fixed-line carriers operate the PHS services. Interestingly, there were more PHS than cdmaOne subscribers in China at this time. Other adopters include Taiwan and Thailand. Kyocera, which was the main investor when DDI Cellular was formed in the mid-1980s, and Carlyle acquired the largest PHS service provider, DDI Pocket, in June 2004 in order to promote further sales of equipment.

PHS would probably have been more successful if the standard-setting process had been open to international participation from the beginning. The openness of the standard-setting process to all Japanese manufacturing firms and service providers from the beginning has caused a different service provider (DDI Pocket) and manufacturers (not NTT DoCoMo's major suppliers) to become the market leaders in Japan. However, the MPT delayed the opening of the standard to foreign firms in order to give Japanese manufacturers an advantage, a move similar to the behaviour often attributed to Japan's Ministry of International Trade and Industry.[31] Foreign firms were not invited to participate in the standard-setting process and the system specifications were not made public until one year after service was started, in July 1995. Quite naturally, most foreign manufacturers, including industry leaders such as Ericsson, Nokia and Motorola, have criticized PHS. This has made it difficult for Japanese manufacturers to persuade foreign carriers to adopt PHS, particularly when many of these foreign carriers are major customers of firms such as Ericsson and Nokia.[32]

Domestically, perceptions about PHS rose and fell several times. The heavy media coverage during the service start in mid-1995 caused many Japanese to visit stores with the intention of subscribing to PHS, but the initial poor coverage caused many of them to subscribe to PDC services. Improvements in coverage led to a short-term boom in subscribers in 1996 that also fuelled heavy competition between PHS and PDC services. For example, the PDC service providers were spending more than ¥70,000 ($580) to acquire a subscriber in the mid-1990s,[33] levels that have never been reached in Europe or the United States.

On the other hand, this competition led to lower PDC rates, thus causing the price advantage of PHS to slowly disappear and the dependence on NTT's wireline system to become a major problem. The PHS systems depended on NTT not only for connections to fixed-line and PDC systems, but also for connecting base stations within a PHS to PHS phone call. NTT, which had a monopoly on local telephone calls throughout the

1990s, set very high connection charges for the PHS service providers and there have been constant battles between the NTT and the PHS services providers over these charges. For example, 40 per cent of the revenues for the PHS carriers were being paid to NTT in the first two years of the PHS service (only 9 per cent for cellular firms), while these connection charges have represented a significant fraction of NTT's profits.[34]

Furthermore, the PHS carriers depended on NTT for fundamental changes in PHS, and NTT has implemented these changes very slowly. For example, as the penetration rate for non-PHS mobile phones grew from about 3 per cent in April 1994, when PHS was being designed, to over 20 per cent by late 1997, the number of people making calls to or receiving calls from these non-mobile PHS phones increased sharply. However, calls between PHS and non-PHS mobile phones were not possible until mid-1996, and when they became possible, they were nine times more expensive than PHS to PHS or PHS to wireline calls, and three times more expensive than calls made between non-PHS mobile phones. The reason for the delays in making the calls between non-PHS and PHS mobile phones possible and the reason why these calls are still very expensive is that NTT has been very slow to implement appropriate switching equipment, and it is generally believed that the MPT has not set appropriate connection charges.[35]

The number of PHS subscribers had declined to a level of about five million by mid-2004 and it is losing a couple of hundred thousand subscribers a year. If not for the superior data services of PHS, the decline would probably be much greater. The three PHS service providers offer 128,000-a-second services with data cards that are widely used with laptops and personal digital assistants (PDAs). Flat-rate plans are particularly popular. Furthermore, about one-tenth of NTT DoCoMo's (it acquired NTT Personal in 1998) PHS subscribers use the service to monitor the location of phones – for example, parents monitor their children's location.

3G standard setting: NTT DoCoMo becomes a global participant

Continued pressure from the MPT and Europe's slow moves to develop cdma technology provided NTT DoCoMo with the opportunity to create a global 3G standard. Japan's MPT pressured NTT DoCoMo to either create or adopt a worldwide standard, because of the pressure it was receiving from domestic and international manufacturers. Although Japanese phone manufacturers had initially done well in the global analogue market in the 1980s, their shares had subsequently dropped, owing to their own mistakes,[36] differences in market structure[37] and the use of non-global standards in the Japanese market.

NTT DoCoMo had little interest in adopting another firm's standard, particularly since it had been working on 3G systems, including cdma-based ones, since the early 1990s. These factors caused NTT DoCoMo to focus on creating a global standard through a much more open approach than it had done with PDC and to eventually create an alliance with Nokia and Ericsson and thus with Europe's GSM (Global System for Mobile Communications) community. The Japanese government also pushed for an early start date, which NTT DoCoMo also wanted, because of its desire for additional frequency spectrum. Owing to the high population densities in Japan and the large share for NTT DoCoMo, NTT DoCoMo has experienced shortages of frequency spectrum since the mid-1990s. For example, it was the first service provider in the world to adopt a so-called 'half-rate' for voice, which substantially reduced the quality of voice calls.[38]

European firms were very slow to recognize the superiority of cdma. In the competition between European and US firms, in particular between Ericsson and Qualcomm, European firms had continuously argued that GSM was superior to cdma. However, as cdmaOne began to diffuse in the United States and elsewhere in 1995 and 1996, the possibility that cdmaOne would be chosen by the newly formed UMTS became a large concern to Ericsson and Nokia, which had previously decided not to participate in the cdmaOne infrastructure market. Their late entry would have made it difficult for them to become competitive, partly since they would be paying higher licence fees than the early entrants. Ericsson was the largest supplier of GSM and mobile communication infrastructure overall, while Nokia was the second-largest supplier of GSM infrastructure in the mid-1990s.[39]

Ericsson and Nokia used their influence in Europe and Europe's plan to choose a single standard in late 1997 to negotiate with NTT DoCoMo. They persuaded NTT DoCoMo to adopt the evolution path of the GSM network interface in place of Docomo's proposed ISDN interface, which would enable GSM infrastructure suppliers and service providers to utilize some of their existing technology in third-generation systems. NTT DoCoMo announced the adoption of this technology in March 1997 and Ericsson and Nokia announced their support for NTT DoCoMo's W-CDMA system ('W' stands for wideband) in May 1997. After more than a month of negotiations between Ericsson, Nokia and other manufacturers, the European Telecommunications Standards Institute (ETSI) selected a combination of W-CDMA and TD-CDMA for the 3G standard. The former would be used for outdoor applications and the latter would be used for indoor applications.[40]

ETSI's selection of W-CDMA also led to changes within Japan. It caused Vodafone to acquire a controlling interest in J-Phone as it became clear that Japan would probably be the first country to implement W-CDMA on a wide scale. As part of their implementation of cdmaOne and their desire to obtain 3G licences and frequency spectrum, DDI Cellular, IDO and KDD merged. KDD (international calls) had planned to apply for a 3G licence, owing to the expected international roaming capability of 3G. All three firms concluded that it would be better to apply jointly rather than compete for a 3G licence. For similar reasons, KDDI acquired Tsuka Cellular.

Mobile Internet: NTT DoCoMo begins foreign expansion

The success of NTT DoCoMo's i-mode service enabled NTT DoCoMo to further increase its share of the Japanese market and also to expand its global presence beyond 3G standard-setting. Although it did not adopt WAP (Wireless Application Protocol), which was the perceived global mobile Internet standard at the time, it did use other global standards and foreign technology. It developed compact HTML (c-HTML) and compact MIDI (Musical Instrument Digital Interface), which were mobile versions of the global standards for markup languages (HTML) and karaoke music at the time. It also purchased servers, database equipment and routers from leading US firms.

Key elements of the initial i-mode service included a micro-payment system, packet service and Internet mail.[41] Although KDDI introduced mobile Internet services in April 1999, only two months after NTT DoCoMo did, KDDI did not introduce packet services until December 1999 and, more importantly, a micro-payment system and Internet mail until April 2000. The lack of a micro-payment system delayed the creation of entertainment content, and there were many technical problems to be solved with WAP. While the overall failure of WAP in Western countries allowed individual mobile service providers to move slowly with WAP, KDDI was forced to solve many problems before they were addressed in the various WAP committees. Like NTT DoCoMo and Vodafone, KDDI was helped by its close relationships with phone manufacturers.[42]

Vodafone introduced its mobile Internet services, which were based on Mobile Mark-up Language (MML), in December 1999, including micro-payment services. Many people believe that the similarities between MML and c-HTML simplified the introduction of Vodafone's services. Although Vodafone did not introduce a packet system until 2001, it charged by the packet. Thus, users did not see large disadvantages in comparison to the KDDI system. In fact, Vodafone and KDDI experienced very similar levels of growth in their mobile Internet services in 2000 and 2001.

Network effects[43] and its early lead enabled NTT DoCoMo to begin building a network of content providers, mobile Internet subscribers and mobile Internet-compatible phones before the other service providers could do this. Within one year of the start of services, there were more than three million subscribers, 300 content providers on the official menu (four times as many as when the services had started) and more than 5,000 content sites that could be accessed via the input of a URL address. The success of entertainment contents, particularly images, ringing tones and horoscopes, caused phone manufacturers to release phones with colour displays and Mobile Information Device Profile (MIDP) capabilities in late 1999 (two phones) and early 2000 (two phones). The high activation commissions paid by NTT DoCoMo and the other service providers contributed to this growth in mobile Internet subscribers since they reduced the cost of technologically sophisticated phones to the final users.[44]

However, technological change appears to have at least partially negated NTT DoCoMo's advantage in network effects. For example, NTT DoCoMo initially turned down Sharp's proposal to produce camera phones. Instead, Vodafone was the first service provider to offer camera phones and a service for attaching pictures to e-mail called *Sha-mail*. Camera phones were a huge hit in spite of the fact that few people actually takes photos with them or attach photos to mail messages. Nevertheless, the success of *Sha-mail* enabled Vodafone to increase its share of new subscribers in 2001 (see Figure 3.5).

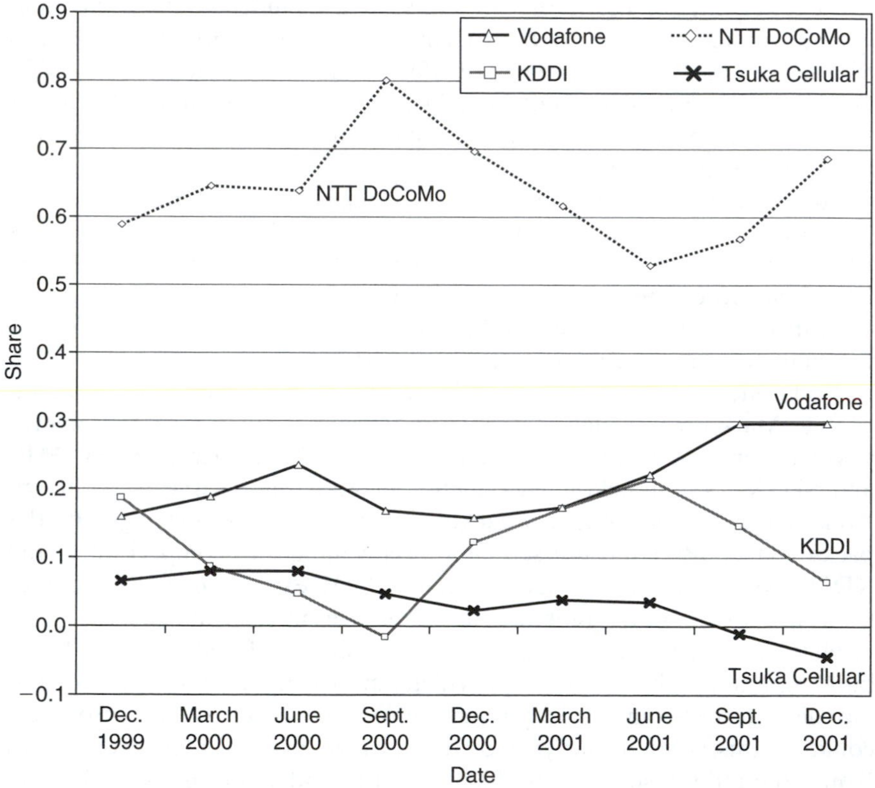

Figure 3.5 Different service providers' shares of net additions in the previous three months, 1999–2001 (source: Japan's Telecommunication Carriers Association and author's analysis).

NTT DoCoMo's foreign expansion

The success of i-mode and the failure of WAP outside of Japan and Korea provided NTT DoCoMo with a global opportunity. While phones for European WAP services were incapable of displaying content in a consistent manner across different phones, NTT DoCoMo's close relationships with phone manufacturers and its close control over the i-mode standard resulted in consistent and high-quality content. The i-mode service also included Internet mail, which is superior to SMS (short message service) services and was still not available in most services outside Japan as of late 2004 (some in the United States). The use of c-HTML and Internet mail services facilitated the entry not only of content providers, but also of many other firms such as retailers, restaurants and manufacturers.[45]

By early 2000, NTT DoCoMo was actively selling i-mode overseas. Although it initially attempted to profit from i-mode via investments in foreign carriers such as KPN and AT&T Wireless, the collapse of the Internet bubble and NTT DoCoMo's lack of foreign experience has caused it to focus on licensing i-mode. The first services were launched by KPN Mobile in the Netherlands and Belgium and KPN Mobile's subsidiary E-Plus in Germany in 2002. Via licences, i-mode was launched by Bouygues Telecom in France and Far EasTone in Taiwan in 2002, by Telefónica Móviles in Spain and Wind in Italy in 2003, and by COSMOTE in Greece in June 2004. These i-mode services had acquired three million subscribers by the end of June 2004. Telefónica Móviles, which is the largest service provider to adopt i-mode, plans to offer services in South America, and Cosmote plans to offer the services in Bulgaria. Extrapolating from the recent growth suggests that the number of subscribers may pass the ten million mark before the end of 2005.

The biggest challenge facing i-mode outside Japan is network effects. Since Europe uses GSM and not NTT DoCoMo's proprietary PDC standard, different phones have to be created for the European i-mode market, which are also different from regular GSM phones. There were only two phones available at the end of 2003, and although four new phones were released in 2004, this number is still far less than the number of phones available with regular GSM services. The lack of phones is said to be one of the main reasons why AT&T Wireless has pushed m-mode in the United States, which requires fewer changes to the basic GSM phones and thus they are easier to obtain than i-mode phones. Furthermore, the incompatibilities between Internet mail and SMS make it difficult for i-mode users to benefit from the existing network of SMS users.

It is interesting to speculate on what would have happened if NTT DoCoMo had adopted GSM in the early 1990s. Not only would this have provided Japan's manufacturers with greater global opportunities,[46] it would have also have made it easier for NTT DoCoMo to license i-mode. Of course, the adoption of GSM would have completely changed the competitive dynamics of the Japanese market, thus raising the possibility that i-mode would not have emerged, or would have emerged in a completely different context.

KDDI's recovery

KDDI's faster diffusion of 3G phones and applications enabled it to obtain more than 50 per cent of the new subscribers in 2003; this was the first time a firm other than NTT DoCoMo had achieved this in Japan (see Figure 3.6). The reason for the faster diffusion of 3G phones for KDDI is the backward compatibility of KDDI's 3G system. While NTT DoCoMo has had to introduce completely new base stations and handsets, KDDI has only had to make changes to the software in the base stations and phones. Thus, while KDDI had achieved more than 99 per cent coverage by April 2002, or within a few months of starting services, NTT DoCoMo did not achieve this level of coverage until early 2004. Furthermore,

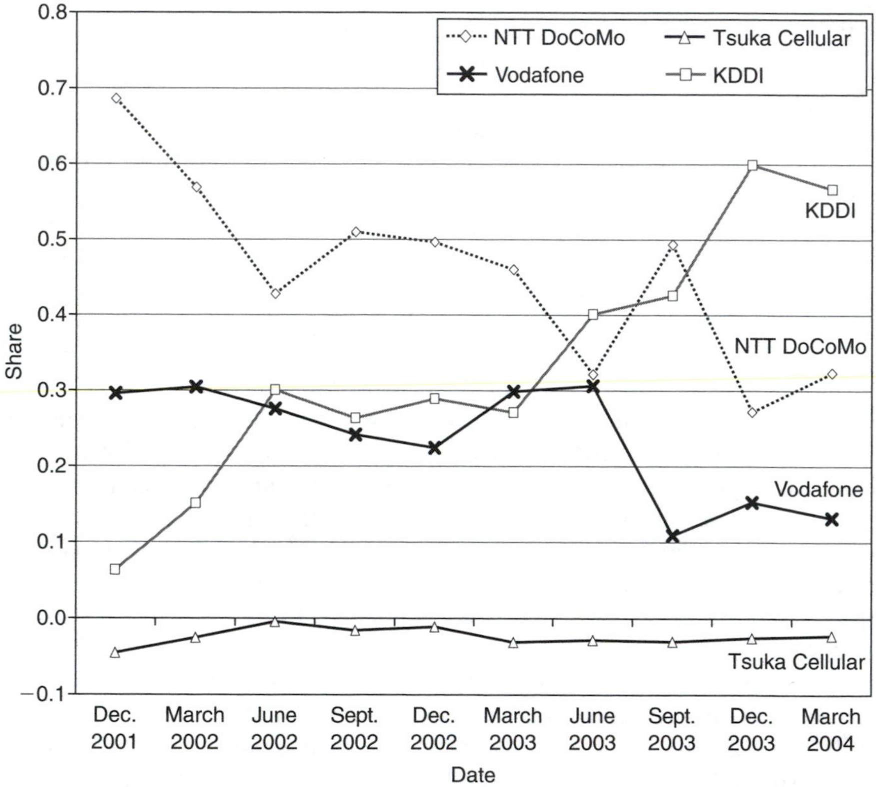

Figure 3.6 Different service providers' shares of net additions in the previous three months, 2001–2004 (source: Japan's Telecommunication Carriers Association and author's analysis).

FOMA handsets are more expensive to users than cdma2001x phones in spite of the fact that NTT DoCoMo has been spending about ¥55,000 on handset subsidies, or more than ¥20,000 more than KDDI has done.

As shown in Figure 3.6, the first rise in KDDI's share came from the introduction of cdma20001x in April 2002. The faster data speeds and higher capacity of cdma20001x also enabled KDDI to introduce a new pricing plan called *Packet Wari* (Packet Discount). This plan offered 12,000 packets for ¥1,200 or ¥0.1 per packet, which was half the price of the regular EZ Web Service and one-third the price of i-mode.

The introduction of *Chaku Uta* led to the second rise in KDDI's share of new subscribers. This new application is a major reason why KDDI's implementation of cdma20001x in 2001 has caused a long-term increase in subscriber shares while its implementation of cdmaOne in 1998 did not. The *Chaku Uta* service enables users to download a 15- to 30-second MP3 song, including lyrics, and use the song as a ringing tone in place of the MIDI files. The number of downloads jumped from six million songs in September 2003 to 50 million in the second quarter of 2004, or 13 million per month; interestingly, this is similar to the number of global downloads with Apple's i-Tune service.

Conclusion

This chapter has looked at the interaction between globalization, deregulation and technological change in the Japanese mobile phone industry. Domestic concerns and both halves of globalization (foreign pressure and the desire to participate in the global market) have driven deregulation in the Japanese market. The US government pressurized the Japanese government to increase competition in the market by allowing the use of foreign technology and foreign investments. Results of this foreign pressure include DDI Cellular's adoption of TACS, DDI Cellular and IDO adoption of Qualcomm's cdma technology, investments by many foreign firms in Japanese service providers, and NTT DoCoMo's steadily increasing purchases of foreign technology throughout the 1990s.

On the other hand, many firms' desire to participate in the global market has also encouraged the Ministry of Posts and Telecommunications to propose these changes. MPT created an open standard-setting process, at least from the domestic standpoint, for PHS in the hope that Japanese firms could export PHS to the rest of the world. The failure of PHS at the global level and the failure of Japanese manufacturers in the global phone market in general caused MPT to require NTT DoCoMo to either create or adopt a 3G standard.

Technological change provided new entrants, which were the result of deregulation, with opportunities, and encouraged domestic firms to use foreign technology (i.e. globalization). TACS provided DDI Cellular and digital technology provided Digital Phone and Tsuka Cellular with an opportunity to enter the market and compete with the incumbents on a 'relatively' level playing field. While Toyota perceived that it was less risky to adopt its competitor's standard than adopt a global standard in IDO's service, the successful adoption of TACS by DDI Cellular in the late 1980s caused it to realize by 1995 that the opposite was true. Its merger with DDI Cellular and KDD has made KDDI one of the world's leading users of cdma technology.

DDI Cellular's success with foreign technology is also one reason why NTT DoCoMo began working with foreign firms and eventually learned to combine these technologies in unique ways. NTT DoCoMo has gradually moved from an internal to an external focus as it increased the use of global technology, and it is now a provider of technology on the global level for the mobile Internet. On the other hand, the increasing strength of the Japanese market attracted the world's largest service provider, Vodafone, to the Japanese market. Vodafone's acquisition of J-Phone and the use of J-Phone's mobile Internet technology is one reason why Vodafone is now the leading provider of mobile Internet services outside Japan and Korea with its Vodafone Live! services.

Further research should look at other Japanese industries and compare them to Japan's mobile phone industries. Understanding why some industries are able to change faster than others is an important subject to researchers of Japan and other countries.

Notes

1 Joel Baum and Olav Sorenson, 'Geography and Strategy', *Advances in Strategic Management*, Vol. 20 (New York: Elsevier, 2003).
2 For example, see Bruce Kogut, 'The Internet Has Borders', in Bruce Kogut (ed.) *Global Internet Economy* (Cambridge, MA: MIT Press, 2003) or Niko Waesche, *Internet Entrepreneurship in Europe: Venture Failure and the Timing of Telecommunications Reform* (Cheltenham, UK: Edward Elgar, 2002).
3 Kogut, 'The Internet Has Borders'.
4 Jeffrey Funk, 'Standards, Dominant Designs and Preferential Acquisition of Complementary Assets through Slight Information Advantages', *Research Policy* 32: 1325–1341 (2003) [hereafter, Funk, 'Dominant Designs']; Jeffrey Funk, *Global Competition between and within Standards: The Case of Mobile Phones* (London: Palgrave, 2002) [hereafter, Funk, *Global Competition*].
5 For example, see Lonnie Carlile and Mark Tilton, *Is Japan Really Changing Its Ways? Regulatory Reform and the Japanese Economy* (Washington, DC: Brookings Institution, 1998); Frank Gibney (ed.), *Unlocking the Bureaucrat's Kingdom: Deregulation and the Japanese Economy* (Washington, DC: Brookings Institution, 1998), or Edward Lincoln (ed.), *Troubled Times: U.S.–Japan Trade Relations in the 1990s* (Washington, DC: Brookings Institution, 1999).
6 Funk, *Global Competition*.
7 Jeffrey Funk, *Mobile Disruption: The Technologies and Applications Driving the Mobile Internet* (New York: John Wiley, 2004).
8 Jeffrey Funk, 'The Product Life Cycle Theory and Product Line Management: The Case of Mobile Phones', *IEEE Transactions on Engineering Management* 51 (2): 1–11 (2004); Funk, 'Dominant Designs'.

9 Jeffrey Funk, 'Mental Models and the Startup Problem: The Case of the Mobile Internet', Hitotsubashi University, Institute for Innovation Research, Working Paper (2004).
10 Philip Anderson and Michael Tushman, 'Technological Discontinuities and Dominant Designs: A Cyclical Model of Technological Change', *Administrative Science Quarterly* 35: 604–633 (1990); Michael Tushman and Philip Anderson, 'Technological Discontinuities and Organizational Environment', Administrative Science Quarterly 31: 439–456 (1986).
11 Peter Grindley, *Standards, Strategy, and Policy: Cases and Stories* (Oxford: Oxford University Press, 1995); Carl Shapiro and Hal Varian, *Information Rules* (Boston: Harvard Business School Press, 1999).
12 Garry Garrard, *Cellular Communications: Global Market Development* (Boston and London: Artech House, 1998).
13 Funk, *Global Competition.*
14 Ibid.
15 Hiro Wakabayashi, 'Japan's Revolution in Wireless', Nomura Research Institute Quarterly (Autumn 2004).
16 Funk, *Global Competition.*
17 Mayumi Shigetaka, 'Keiei senryaku – gosan no kenkyu – nihon ido tsushin – sanju toshi de shueki akka NTT hoshiki saiyo ga ashikase' (A mistake in creating a management strategy by Japan's IDO – three investments including NTT's system are holding back profits), *Nikkei Business* 30 (January): 43–45 (1995); Laura Tyson, *Who's Bashing Whom: Trade Conflict in High Technology Industries* (Washington, DC: Institute for International Economics, 1992).
18 Funk, *Global Competition.*
19 Shigetaka, 'Keiei senryaku', Tyson, *Who's Bashing Whom.*
20 Funk, *Global Competition.*
21 Ibid.
22 Ibid.
23 Funk, 'Dominant Designs'; Funk, *Global Competition.*
24 Funk, 'Dominant Designs'.
25 Ibid.
26 Ibid.
27 Ibid.
28 Wakabayashi, 'Japan's Revolution'.
29 Toshihiro Tonishi, 'Analysis: Kyocera Going Global with PHS, Nikkei Online', http://neasia.nikkeibp.com/wcs/leaf/CID/onair/asabt/news/315644 (June 2004).
30 EMC Report, 'PHS Set to Enjoy Continued Growth in China', http://wcis.emc-database.com/marketdata.nsf/popupdocs/422039D912E5F6D780256F04006A5A92 (June 2004).
31 Chalmers Johnson, *MITI and the Japanese Miracle, Tokyo: The Growth of Industrial Policy, 1925–1975* (Stanford, CA: Stanford University Press, 1982); Clyde Prestowitz, *Trading Places: How We Allowed Japan to Take the Lead* (New York: Basic Books, 1988).
32 Jeffrey Funk and David Methe, 'Market and Committee Based Mechanisms in the Creation and Diffusion of Global Industry Standards: The Case of Mobile Communications', *Research Policy* 30 (4): 589–610 (2001).
33 Funk, *Global Competition.*
34 Ibid.
35 Ibid.
36 Funk, 'Mental Models'.
37 Funk, 'Dominant Designs'.
38 Jeffrey Funk, 'Competition between Regional Standards and the Success and Failure of Firms in the World-Wide Mobile Communication Market',

Telecommunications Policy 22 (4): 419–441 (1998); Funk and Methe, 'Market and Committee Based Mechanisms'.

39 Funk and Methe, 'Market and Committee Based Mechanisms'.

40 Ibid.

41 Funk, *Mobile Disruption*; Funk, 'Mental Models'.

42 Funk, *Mobile Disruption*; Funk, 'Mental Models'.

43 Shapiro and Varian, *Information Rules.*

44 Funk, *Mobile Disruption.*

45 Ibid.; Funk, 'Mental Models'.

46 Funk, *Global Competition.*

4 Changing satellite systems in Japan within a global context

Ruth Taplin

Satellite communications are recognized by the Japanese as being an essential infrastructure to facilitate wide coverage, flexible networking, dealing with and preventing disasters, and many other functions. Although the Japanese are developing their own satellite systems, which are becoming more user-friendly, many problems for the technical development of satellites continue to exist such as creating a seamless service that is linked to the terrestrial system, or the scope and breadth of global positioning systems (GPS). GPS, because of their capacity to be used for politically sensitive areas such as defence and intra- or inter-country surveillance, have given rise to geopolitical debate in terms of their development, especially in relation to the Galileo GPS, which is drawing Japan into a global competition that it is reluctant to be a part of.

Japan is determined to develop its own satellite system and is combining the efforts of research and development (R&D) organizations, administrative agencies and the space industry to bring new developments to fruition, especially the Quasi-Zenith Satellite System (QZSS), the next generation of low earth orbit (LEO) satellites and milliwave/optical satellite.

The Communications Research Laboratory (CRL) of Japan was attempting to develop a satellite that can be used anywhere, by anyone and at any time, along with a large-scale system such as a geostationary orbit platform. CRL's future activities will now be carried out by the National Institute for Information and Communications Technology (NICT), including an optical satellite that may be integrated with a stratosphere platform, a smart satellite that can demonstrate new technology on orbit in a short period, basic research to monitor and detect satellites in orbit or space debris from the ground, and new technology development such as large-scale phased array antennas. NICT has many plans to move forward development of GPSs, which will be described further on in this chapter.[1]

The Japanese Quasi-Zenith Satellite System in development

From the Japanese perspective, the QZSS is a potential panacea that will create new markets, generate income and employment, and bring hitherto unavailable social benefits. It is estimated that the direct economic benefit could total ¥0.3 trillion ($2.7 billion) and that including related markets, service, terminal and satellite systems markets, the total effect over a period of 12 years will be worth ¥5.7 trillion ($52.1 billion). The economic knock-on effect from these markets will result in generated income of ¥21 trillion and the creation of 810,000 jobs. If the GPS applications are added, the estimated social benefit from QZSS will be ¥2.4 trillion.

There exists incontrovertible evidence that the possibilities are there, but Japan cannot operate in isolation, and the direction of the American GPS, the European Galileo project and the role of China will all affect the outcome. The rudiments of QZSS are sound, however.

The basic principle of QZSS is to create new business and social opportunities that will benefit the Japanese economy and society through multiple broadcasting/communications/positioning. QZSS aims to expand GPS satellites' positioning by transmitting the same positioning signals as the 24 GPS satellites developed in the United States, which are relied on globally. To compensate for positioning satellite errors that allow interference from mountains and buildings to obstruct signals, QZSS will involve positioning three satellites in orbit that will alternate sequentially in the skies over Japan every eight hours. In practice, this will mean that there is always one satellite in Japan's zenith area. Positioning three satellites on orbital planes separated by 120 degrees and synchronizing their orbital paths enables a 24-hour unobstructed view of satellites from Japan from an angle of elevation of 70 degrees and over, which will allow a high standard of mobile communications with minimal line interference. A single beam will be required to provide a uniform service across the country.[2]

With regards to high-speed mobile communications, the satellites will only have relaying capabilities and scaled-down antennas with a Ku-band transponder. To provide greater accuracy and delivery of signals such as compensation signals to users, alternative means will be utilized to shorten positioning time. Like the European Union's Galileo system, QZSS will be attempting to reduce the margin of error due to ionospheric conditions which cause interference to transmission signals to the users. QZSS will monitor such margins of error simultaneously from nearby positioning control points that can compensate for positioning errors by transmitting the volume of errors to users at appropriate times. Work on the first satellite was begun in fiscal year (FY) 2004 with a view to launching it in mid-2008, while satellites 2 and 3 will be launched in the mid to latter part of 2009.

To make the QZSS system commercially viable, the Advanced Space

Business Corporation (ASBC) and NICT are looking at ways to simplify the satellites while increasing their capabilities. In terms of broadcasting needs, for example, terminals being used on moving vehicles need to be scaled down to as small a size as possible.

As the chapters in this book note, Japan was initially slow to embrace the Internet and the digital age but is now pushing ahead with building the infrastructure for terrestrial fibre-optic networks, cell phones, terrestrial digital broadcasting and positioning services. Takanori Ida in Chapter 2 provides evidence of the remarkable recent success Japan has had in developing broadband services in Japan, accelerating access to a wide range of Japanese customers at some of the most competitive prices in the world.

In-car navigation services and sensors based on GPS are a standard feature now of all Japanese cars and major road systems. However, despite this, current GPS satellites are not accurate enough in timing and positioning, with resulting problems such as major differences in accuracy and stability depending on the time of day, and they have difficulty penetrating sea beds. QZSS would not, however, replace the current GPS system at present emanating from the United States but would coexist alongside it, resolving these current infrastructural problems while expanding areas that require coverage.

The geopolitical framework

Global positioning applies not just to the technical aspects of satellite development and GPSs, but also to the competitive positioning of the world's major and upcoming powers, which view being in the forefront of satellite development as key to future economic, political and social success. The United States is concerned that its pioneering GPS system, originally developed and still controlled by the military, is becoming out-of-date technology. Japan, the world's second largest economy, wants to be in control of its own satellite system and have its economy reap the benefits, while at the same time not to offend the world's only remaining superpower, the United States. Therefore, the QZSS emphasizes its interoperability with the US system and looks to the future working with American GPS developments in 2010, when the next-generation satellites will be developed alongside a terrestrial control network. Russia is launching its own GPS named GLONASS by 2011 but this does not appear to be favoured in use by Japan, the United States or Europe. In 2003, China launched its third positioning satellite, called the Beidou Satellite, and has also invested substantively in 2004 in the European Galileo project. Galileo, which is currently being built, is a major competitor to all the GPS systems, as its applications are the most advanced in the world and it will be civilian operated.

The major players in GPS competitive development at present are the

United States and the European Union, and the outcome of this competition will profoundly affect not just Japan but the world. Should the United States and Europe work together to ensure interoperability between the GPS and Galileo, the latter would improve both the quality of satellite navigation and timing services for all users globally. If they do not work together gainfully, a split for competing interests could occur, with countries like Japan making decisions based on loyalties and political expediency rather than on need. The US GPS system would be inferior to the Galileo, causing the US military to invest large amounts of money to try to reach the standard of Galileo. There is no reason scientifically why the two systems could not have interoperability whereby a mutually appropriate design was created that would allow for the minimum interference between the two systems, reducing the interference level to below 0.3db even in emergency usage.

At present, the depth of cooperation between the European Union and the United States regarding the next generation of GPS is not sufficient to create interoperability. A basic obstacle to increased cooperation is not only who will accrue the economic benefits from such developments, but that GPS is run by the US military while, as mentioned, the EU Galileo will be a civilian-run operation. Yet if the cooperation were carried out properly, there is little reason for it to affect US military interests adversely, and it could lead to more reliable signals for civilian services within the US Defense Department. There is also the possibility that if the Pentagon were able to retain total control over military signals, more positive discussion would emerge about the integration of civil applications of GPS and Galileo under international management.

Galileo to date is superior to any other GPS in the world today. It will provide ten signals for civilian use by 2008, making it more reliable, while GPS intends to offer only three signals by 2015. Galileo also has the advantage that many EU states contributed to its creation under the European Space Agency, which means that there is a likelihood that all EU member states will wish to use Galileo rather than GPS. China has made a substantial contribution to the Galileo programme and will therefore be compelled to use it. US companies might find it more difficult to use or buy into Galileo because of EU policies and US export restrictions.[3]

Galileo will be a civilian system that operates for profit, charging its users. It is not clear that such business opportunities would wish to be shared with another system. At present, it appears that GPS, Galileo and GLONASS will become regional systems, with Japan, India and Brazil major players in the GPS camp. However, if GPS fails to deliver the advanced services of Galileo, Japan may be compelled to buy into the EU-based services.

This issue holds such great importance because whoever rules the global satellite infrastructure has the potential to rule the world. The geopolitical alliances are crucial therefore to a balanced power structure in terms of ownership and access to vital information for defence systems, remote location and information networks. China, for example, is at present in the Galileo camp, having invested in it. India has a vested interest in the GPS system. However, should the two giants of China and India, which hold the bulk of the world's population decide to join forces and control or develop an even more advanced system jointly, distribution of power globally could become imbalanced. Given the recent friendly overtures and signing of mutually beneficial trade agreements concerning the electronics industry, the possibility of such joining of forces with regards to satellite development is not simply speculation. The same scenario could hold true if the United States, Europe and Japan all decided to use and further develop one system, especially if it was to the exclusion to the rest of the world. Less advanced nations are already behind in control of the world's resources, information, intellectual property, corporate strength, and so forth. Should they not have equal access to, or access of some dimension of, these powerful information-gathering satellite tools, they would fall even further behind into an exclusion zone. When Galileo GPS, for example, begins charging for services, some of the costs may be beyond the reach of less advanced nations, which will put them at an immediate disadvantage. The Galileo system will have to develop a very strict code to ensure that its products are readily available to all while at the same time not misused by a nation or individual with unsavoury intentions.

What is Galileo and why is it so competitive?

The new European Galileo Global Positioning System (GGPS) not only offers the most advanced scientific possibilities for GPS in the world, which will be explored further on, but has a political, economic and social imperative. It will help to unify Europe by boosting exports in technology and telecommunications services globally to an estimated value of €20 billion per annum, with the British share estimated to comprise at least 10 per cent of this sum. It will greatly stimulate employment.

By its nature of providing quicker and more accurate information to Europe, it will stimulate overall coordination in Europe and serve to integrate the European Union's newly admitted East European nations. It will also ensure that European defence operations do not become entirely dependent on a military power outside Europe.

The Galileo project, named after the eminent Italian astronomer-scientist, is a plan by Europe to put GPS into the twenty-first century. The GPS system initiated and controlled by the United States, which dominated the world of navigation and tracking in the twentieth century, is now 20 years old. The new GGPS could both update and revolutionize the current system. Galileo is designed to be superior to GPS in terms of performance, reliability, accuracy, integrity and functionality. An essential point is that it will be certifiable for safety and security applications such as aeronautical ones. As a civilian-controlled system, it can be marketed and accepted worldwide without fear of a hidden agenda such as European plans to interfere militarily on a global scale. GGPS will have the edge over GPS, QZSS and GLONASS because as satellite navigation it offers key distinct advantages in relation to terrestrial positioning systems, notably global, homogeneous coverage, including coverage of the deep ocean, at a reasonably low cost.

The new features of the European GGPS include an impressive 34 satellites and a digital signal structure based on modern digitization. The advantage of such digitization is that the digits are difficult to interfere with, making them easy both to protect and to adapt.

One of the unique features of the system is the compensation it provides for ionospheric particles caused by the small explosions that take place on the sun (commonly known as sunspots), which are responsible for many of the communication problems experienced by existing navigation and tracking systems as well as mobile telephones.

The Galileo system offers intensive memory, absolute accuracy for global positioning, and the ability for those receiving signals to respond back to GGPS. This means that a person can be anywhere in the world and find out where they are, which is invaluable for explorers or oil and gas workers who risk being lost in remote areas.

Galileo is so versatile that it can be used for ploughing fields accurately, charting straight lines and then recording where they are on the ground, navigating smart cars from depot to depot, repositioning oil platforms, and relocating through its memory system the exact location of a rewarding fishing ground.

Galileo is as accurate on the seabed of the earth as it is in space. It can be used in Orbit Control Systems, ensuring that several satellites can fly together while maintaining an accurate relative distance to each other.

Remote sensing for satellite or aircraft is another application in which a space-borne Global Navigation Satellite Service (GNSS) receiver is used to collect range data from accurate carrier phase measurements to supply scientific information about the earth such as atmospheric properties or oceanography. On-ground post-processing of remote observations is always required

Another application is precision surveying for hydrographic vessels inshore and offshore. This will offer positional information to determine detailed images of the seabed in ports, harbours, estuaries, coasts and oceans for safe passage and data for dredging applications. GPSS has the capacity to penetrate the seabed with the aid of acoustic sensors. GPS or GNSS signals create a single, coordinated system with the acoustic sensors through the use of transponders. A boat or a buoy, for example, can receive GPS signals above the water; these are then tied to the acoustic sensor under the boat or buoy, making a seamless system for underwater positioning applications. Acoustic sensors use sound like dolphins or whales by emitting signals which bounce back when hitting an object in the water or seabed, then link to the GPS signals above the water to facilitate accurate positioning.[4]

Oil and gas exploration is another area where Galileo can be of use: its enhanced accuracy can provide precise positioning and navigational information for marine vessels that undertake profiling of the seabed geography, identifying geomorphological or geophysical hazards that may impact on drilling activities. Its abilities could also be used to good effect as warning systems for seismic activity of the kind that caused the 9.0 earthquake and tsunami off the coast of Sumatra in December 2004.

Traffic surveillance and monitoring for road surveillance and regulatory enforcement will be an important facet of GGPS for a variety of reasons. One application will allow regulatory authorities to monitor the speed and position of vehicles to ensure that speed limits are not exceeded, while another is to monitor the use of untaxed vehicles.

Fleet management and asset management for the former will record position, speed, heading and time information, reporting to the control centre to improve operations and the management of fleet vehicles, while for the latter it is a portable device for recording the position and time information of fixed assets. The tracking of assets such as containers, ships and boats is also part of this process. GGPS can also be used for non-scheduled aircraft flights carrying passengers, cargo or post. Emergency services, air ambulances and police aircraft will also benefit.

Then there is personal navigation. There are already handheld navigation devices available that use the existing GPS system; they combine map displays and secondary communication functions, and can also be clipped to a bicycle or to the dashboard of a motor vehicle. They currently prove invaluable for hikers, cyclists, golfers and in high-risk sports such as speed-boat racing or motor rallying. Lone worker protection is another application that applies to professional workers supplied with terminals as part of employers' responsibility to provide a safe environment for their employees who work in remote locations. Galileo GGPS will make these devices even more effective. Galileo will also offer a range of chargeable services with additional features such as signal integrity, improved reception, and improved accuracy and availability – all with a guarantee of service delivery.

As mentioned, Galileo will not operate under military control, as the current American GPS system does, but be administered by a European operating company or consortium. In the event of a military crisis, therefore, at least some of the GGPS network will remain operable. Non-military control will make it less likely that all GGPS services could be turned off. A military crisis would be a catastrophe for business, government and institutions throughout the world, because they would no longer have access to all the essential navigational and tracking services provided by GGPS.

The bands available for Galileo download have jumped from 15 to 102 MHz. System design is being finalized now and a decision by the public sector to proceed with it was taken in December 2004. The first prototype Galileo satellite is being built at the University of Surrey, England, which has a reputation for building satellites for countries around the world to spot natural disasters.

Whether Japan, along with other countries in the world, will buy into the superior services offered by Galileo or whether the geopolitical dimension will interfere with this economic process remains to be seen. In the meantime, Japan is continuing to develop its own GPS proficiency for internal needs in particular.

The most recent satellite/GPS developments from Japan

Hiroshi Nishiguchi of the Japan GPS Council, in a recent updated report from Japan, outlined the most recent GPS and satellite developments in Japan.[5]

The services of the Japanese national standard time and frequency are conducted by NICT, which routinely uses Caesium atomic clocks linked to the international atomic scale (TAI/UTC). NICT is a member of CGGTTS,[6] plays a role in the GPS Common View Program, and works for TAI (International Atomic Time) GPS Time Transfer Network organized by the Bureau International des Poids et Mesures (BIPM) as the main node of the Asia and Oceanic region.

Since 10 June 1999, NICT has been providing its dissemination services of the standard time and frequency (Local UTC), through a newly established broadcasting station using 40 kHz long-wave frequency, and then the 60-year-old historical service system using short-wave frequencies was phased out at the end of March 2001. NICT encourages wider utilization of the above new services as well as the use of GPS-based time and frequency, which are expanding into telecommunication business fields in Japan.

GPS fixed reference stations network, GEONET

The Geographical Survey Institute (GSI, a governmental agency), responsible for maintaining the geodetic control point network in Japan, had already completed, by the middle of 1999, the establishment of a network of 947 continuous GPS fixed reference stations throughout Japan, named GEONET (GPS Earth Observation Network). This network is routinely used, as one of the most basic tools, for crust movement monitoring, contributing to the swift restoration countermeasures against natural calamities, and also will be used for surveying and mapping. Further, GSI has been promoting its project to extend GEONET and had expanded up to 1,224 reference station networks by the end of March 2004.

Disaster management

Additionally, in order to make provision for the recent possible risks from volcanic activities and for seismic active events such as the Tokai earthquake swarm zone, these GPS-based reference stations have been reinforced to a total of approximately 1,500 stations, in which mobile GPS observation units with solar panels, wind generators and satellite communication devices are included. The other relevant universities and laboratories also continue to run a number of their own GPS-based and some GLONASS-based reference stations for contributing to seismic active research (see Figure 4.1).

New geodetic system of Japan

These GEONET reference stations meet World Geodetic System (geocentric system) standards, and the GSI has provided transformation parameters from Tokyo Datum, which was the traditional primary geodetic standard, being used in Japan for more than 100 years. As a

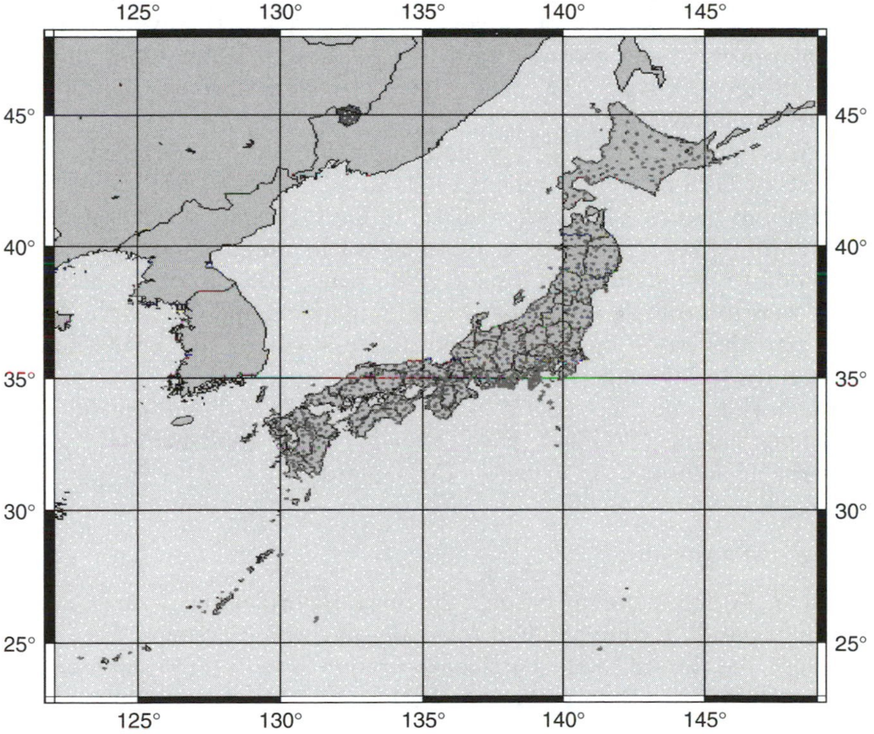

Figure 4.1 Nationwide Global Positioning System (GPS) reference stations in Japan (source: Japan GPS Council).

consequence of the successful GEONET establishment and GPS-based precise monitoring activities for a nationwide land scale, Japan decided to change its geodetic system from Tokyo Datum to the new coordinate system – that is, the ITRF94 standard, which is an internationally common standard. The so-called Geodetic Coordinates 2000 has been enforced since 1 April 2002 as the 'Japanese Geodetic Datum 2000' (JGD2000), which is outlined below.

Referring to global reference frame ITRF94 (International Terrestrial Reference Frame) and GRS80 ellipsoid for geographical coordinates, the geodetic network is based on VLBI and GPS (GPS-based control points). Geoid height data set is provided to determine orthometric height from a GPS observation.[7]

Data from GEONET for post-processing and for real-time kinematics (RTK) positioning

Data obtained by GEONET for post-processing have been provided to researchers and surveyors through the Internet, and currently GSI has added its real-time data services for network-type real time kinematics (RTK) positioning via the server control centre run in cooperation with the Japanese Association of Surveyors. Users such as surveyors and GIS national programmes can now easily access these real-time data, directly or from network-type RTK data service providers, through the links of mobile communication networks.

GSI is implementing a plan to modify a major part of GEONET, which consists of GPS reference stations to RTK data service specification. The 200 stations had already been modified to the RTK data service specification before May 2002, and as of today such modifications have been extended to 931 stations on a nationwide scale. The aim of the series of projects is to facilitate the national-level remapping and GIS implementation activities based on the new coordinate standard 'JGD2000' and to support the 'e-Japan' governance programme.

Meanwhile, the electronic coastal charts (ECDIS) based on the IHO recommendation (WGS84) have already been implemented by the responsible authority, the Japanese Coastguard.

Navigation activities

In Japan, GPS has already become the most important and, indeed, essential information technology tool for many navigation systems, fleet management, the related Intelligent Transportation System (ITS) services, and the location-based information services combined with mobile phone applications.

Augmentation systems and services

There are six kinds of augmentation systems in Japan, for aviation, maritime, car navigation, RTK precision surveying and civil engineering works. Five services, excepting Multifunctional Transport Satellite-based Augmentation System (MSAS), have already been provided.

Car navigation

The Japanese car navigation industry is already one of the most important of all in current GPS utilization, from the economic viewpoint of the various kinds of information technology industry. A nationwide Differential Global Positioning System (DGPS) service using FM subcarriers has been contributing to further growth of car navigation and location-based service business since commencement of its service in May, 1997.

This commercial DGPS service is conducted by the private-sector entity GPex, founded by voluntary support from the 17 members of the Japan GPS Council. GPex operates seven DGPS reference stations throughout the country, and DGPS augmented data are broadcast for public, with DARC-type FM multiplex formula, known as one of the ITU-R standards, via the FM broadcasting network of the existing 43 transmitting stations all over Japan. The DGPS data are interchangeable with the international standard RTCM SC-104 and NRSC/RBDS format in the United States.

The service is provided without direct user fees, and many of the components are designed to make the services for common use to receive the Vehicle Information and Communication System (VICS) services as well. VICS services are also using the same DARC-type FM subcarrier. It may lead to a reduction in the production cost of hardware, and have synergy effects in both marketing and improving ecology. It has been estimated that approximately 1.2 billion litres of fuel can be saved by the effect of popularization of car-navigation products and VICS services in 2010, and it will also correspondingly bring about a huge reduction in CO_2 emissions.[8]

Wide-area augmentation for aviation, MSAS (SBAS)

The Multifunctional Transport Satellite-based Augmentation System (MSAS) represents a GPS augmentation system based on the use of geostationary satellites. The MTSAT (Multifunctional Transport Satellite) will be deployed at the allocated orbit slot, 140° E, with two main missions for enhanced aeronautical services and for continuous regional meteorological services respectively. The Aeronautical Mission provides for aeronautical mobile satellite services for voice and data communication and surveillance (ADS), and also GPS augmentation for satellite navigation aid.

The objective of MSAS is to augment GPS SPS signals sufficiently to support the civil aviation use of GPS, to meet operational requirements for various phases of flight:

- GPS-like additional ranging signals for service availability and continuity;
- integrity information;
- differential correction.

MSAS ground facilities composed of master control stations, ground monitor stations and monitor and ranging stations have already been established. The first MTSAT satellite, however, failed to be launched successfully. The alternative satellite is expected to be launched soon.

The entire national fleets of coastal support vessels, coastguard vessels and lifeboats are fitted with Loran-C receivers. Nonetheless, many of them have already mounted GPS receivers as well: more than 90 per cent of ocean and fishery vessels, and about 25–30 per cent of coastal fishing boats as of today. Then, in compliance with the requirement of maritime beacon DGPS networks to meet the demands of public safety, the Japanese

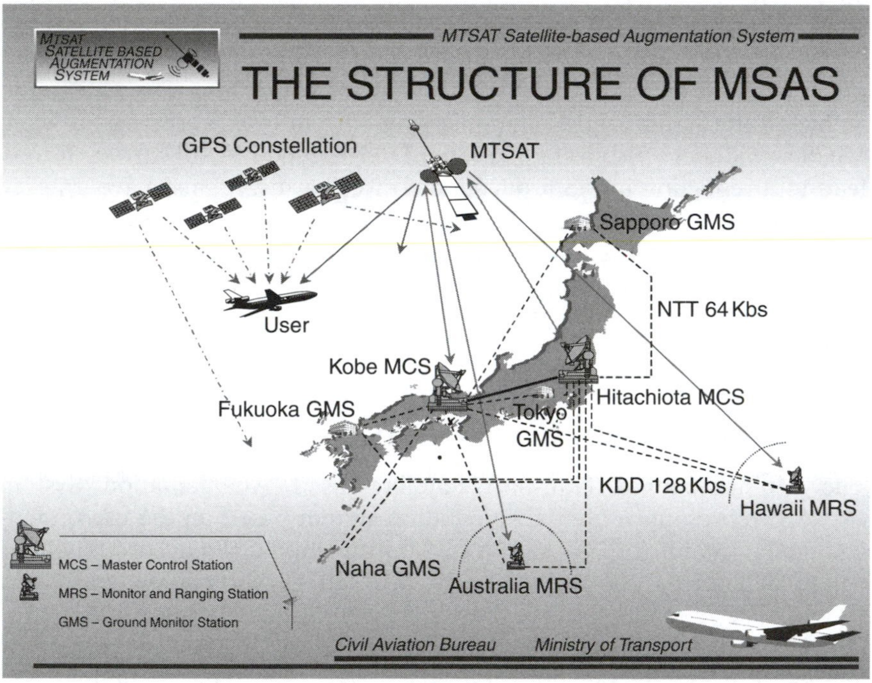

Figure 4.2 Maritime Differential Global Positioning System (DGPS) service for nationwide coastal areas (source: Japan GPS Council).

Coastguard continued its implementation efforts, and at the end of March 1999 achieved full operational capability of the radio beacon DGPS network, with 27 DGPS reference stations in totality covering all the coastal areas of the country.

The DGPS correction data are transmitted via the radio beacon sites listed in Table 4.1 (some of them having the generating of DGPS data as their only function), and the services are non-encoded and without direct user fees. Consequently, the maritime DGPS services will allow remarkable growth in DGPS use in coastal fishing, and also for pleasure boats.

RTK-GPS precision positioning services

RTK-GPS has become popular today in several surveying applications for precise real-time measurements in surveying, construction and machine control applications, etc. Two non-profit organizations and two from the

Table 4.1 List of Differential Global Positioning System (DGPS) stations

DGPS station name	*Latitude*	*Longitude*	*Frequency*	*Transmitting station ID*
Kusiro-Saki	42-58 N	144-23 E	288 kHz	630
Abasiri	44-00 N	144-18 E	309 kHz	631
Soya-Misaki	45-31 N	141-56 E	295 kHz	632
Syakotan-Misaki	43-22 N	140-28 E	316 kHz	633
Matumae	41-25 N	140-05 E	309 kHz	634
Hamada	34-53 N	132-02 E	305 kHz	635
Tango	35-44 N	135-05 E	316 kHz	636
Hegura Sima	37-51 N	136-55 E	295 kHz	637
Sakata	38-57 N	139-50 E	288 kHz	638
Siriya-Saki	41-26 N	141-28 E	302 kHz	639
Kinkasan	38-17 N	141-35 E	316 kHz	640
Inubo-Saki	35-42 N	140-52 E	295 kHz	641
Urayasu	35-37 N	139-54 E	321 kHz	642
Turugi-Saki	35-08 N	139-41 E	309 kHz	643
Hatizyo Sima	33-05 N	139-51 E	302 kHz	644
Nagoya	35-02 N	136-51 E	320 kHz	645
Daio-Saki	34-16 N	136-54 E	288 kHz	646
Muroto-Saki	33-15 N	134-11 E	295 kHz	647
E Saki	34-36 N	135-00 E	320.5 kHz	648
Ohama	34-05 N	132-59 E	321 kHz	649
Seto	33-26 N	132-13 E	320 kHz	650
Wakamiya	33-52 N	129-41 E	295 kHz	651
Ose-Saki	32-37 N	128-36 E	302 kHz	652
Toi-Misaki	31-22 N	131-20 E	309 kHz	653
Tokara Nakano Sima	29-49 N	129-55 E	320.5 kHz	654
Gesasi	26-36 N	128-09 E	288 kHz	655
Miyako Sima	24-44 N	125-26 E	316 kHz	656

private sector have already been established to provide GPS observation data for RTK positioning applications.

RTK-GPS service for land survey using MCA radio channel

One is the services by the RTK-GPS Experiment Promotion Association for encouraging the effective use of RTK-GPS technology to users of general construction, general surveying and increasing RTK-GPS applications. This organization was established in July 1997 through the co-partnership of the Japanese Association of Surveyors with the National Mobile Radio Centres Council, which operates multi-channel access (MCA) radio communication systems. The association now provides for all types of demand in relation to transmission services for all one-second carrier phase data from specifically arranged 22 GPS-based control stations, through MCA mobile radio channels using both 800 MHz analogue and 1.5 GHz digital.

It is programmed for effective use of the existing MCA mobile radio network conducted by the National Mobile Radio Centres Council and the nationwide GEONET conducted by GSI. It is also expected to achieve increasing productivity and economical effectiveness, and consequently to expand various kinds of applications to meet local and regional needs for precise positioning.

DGPS and RTK-GPS service for marine construction

The other is the dedicated service for facilitating offshore RTK and DGPS utilization in offshore construction and civil engineering works. This organization has been established as a closed promotional body by a marine construction group composed of 36 members. It has RTK reference service stations in six major bay areas in Japan, and provides its members companies with pay services of both DGPS and RTK-GPS using a specifically allocated 400 MHz spectrum.

Network-type RTK services

Currently, network-type RTK applications are viewed as a new augmentation technology by virtual reference station networks (VRNs). In Japan, GSI and the surveying industry jointly conducted successful VRN-RTK realization experiments. This led to two private-sector service providers being established based on the Virtual Reference System Network RTK technology. These service providers make use of the real-time data services of GEONET.

Thus, Japan has been using DGPS and RTK-GPS effectively for increasing the productivity and efficient working of general and offshore construction.

Table 4.2 Outline of differential GPS and relative kinematics (RTK) services

Ref. station	*Operator*	*Distribution*	*Charges*	*Users*
27 stations	Japan Coastguard	Maritime radio beacon	Free	Coastal public safety
Beppu City	Fishermen's Union	Marine Phone	Closed service	Fishermen's Union
7 stations	Private sector Gpex	FM subcarrier	Free	Car-Navi users
MSAS	JCAB	MTSAT satellite	Free	Under construction
6 counties	A RTK-GPS	MCA and D-MCA	H/W rental fee	Land surveyors
Promotion body	Mobile radio channel			
Main ports and coastal area	A promotion body A dedicated channel	Pay service	Marine constructions	
		400 MHz band		
GEONET	Private sectors	Mobile phone links	Pay service	Surveying, GIS, etc.

Source: Japan GPS Council.

Development activities by governmental agencies and related industries: facilitating GPS uses for survey, geodesy and Geographic Information Systems (GIS)

As stated earlier, GEONET conducted by GSI continues to grow, and many of the GEONET sites have been implementing modification programmes available to extend RTK-GPS and VRN real-time data services to other communication media such as mobile phone links. Following a successful experiment using VRN technology during 2000 and 2001, the server service centre that controls GPS data services in various ways has been established and commenced VRN data services in May, 2002. Currently, modified GEONET sites have been expanded up to 931 stations and further implementation will be continued.

Educational programmes for switching from the conventional triangular net survey systems to the GPS-based survey systems are now being implemented as a long-term programme by the government (GSI and the Ministry of Land, Infrastructure and Transport, MLIT).

Standardization activities for the GIS format are being progressed through the cooperation of all of the ministries and agencies within the leadership of MLIT, which is one of the important 'e-Japan' programmes.

Automated and remote controlled construction works

Activities to develop remote-controlled operation systems for construction machinery are vital for field applications in dangerous zones such as areas of volcanic activity. These systems are usually combined with RTK-GPS positioning, CCTV graphical data network and computer-aided CAD systems. A new market is emerging for construction machinery mounted with GPS receivers and antennas.

Precision farming

Long-term R&D activities for precision farming take place through the authorities of agricultural R&D institutes, universities and related industries. Related farming machinery manufacturers are marketing their products mounted with dual-frequency GPS positioning systems.

Maritime and water front control

Hydrographic surveying and river control administration are under the control of the Japanese Coastguard and MLIT.

Aviation use

The LAAS programme within Ground-Based Augmentation Systems (GBASs) was offered by JCAB to develop and evaluate Airport Pseudolite Services in 1998, and the evaluation test of the experimental system was implemented in 2001. Assessment of these achievements and future policies is continuing.

Space use

R&D and experimental activities for the improvement of the launch vehicle control, satellite autonomous control on orbit, and precision rendezvous technique in space using GPS have been implemented by the Japan Aerospace Exploration Agency (JAXA) and the space industry.

Precision meteorology

The joint R&D activities for precision meteorology with GPS by and among the Japan Meteorological Agency (JMA), GSI, universities and research institutes are important for Japan. The data from GEONET continue to contribute to these R&D activities as well. Precise meteorological data are becoming partially available for business applications. Local weather forecasts such as the thunder observation network services are one of the emerging business applications.

Plan for a regional navigation and information satellite system (QZSS)

JAXA (formerly NASDA) and NICT (formerly CRL) have been developing basic technologies for satellite-based positioning, including the satellite-mounted atomic clock, constellation time transmission control and high-precision orbit determination. Recently, debates on possible implementation and operation of the QZSS for regional complementariness and augmentation of the GPS system, and developmental conceptions, have been taking place in Japan, as the civil use of GPS progresses rapidly. These debates are becoming more heated, centring on political, administrative and space industry circles, subsequent to the Volpe Report on the GPS Vulnerability Assessment (October 2001) and the political decision on the promotion of the Galileo project in the European Union, and the formation of a cooperative group for realization of the QZSS has been progressing.

Industrial aspects: car navigation

The car navigation market in Japan continues to grow. The markets have been growing year on year since the first GPS-based car navigation product was placed on the market in 1991. The sales volume recorded close to two million units during 2000, when DVD display gained popularity and accelerated the expansion of the car navigation market through the

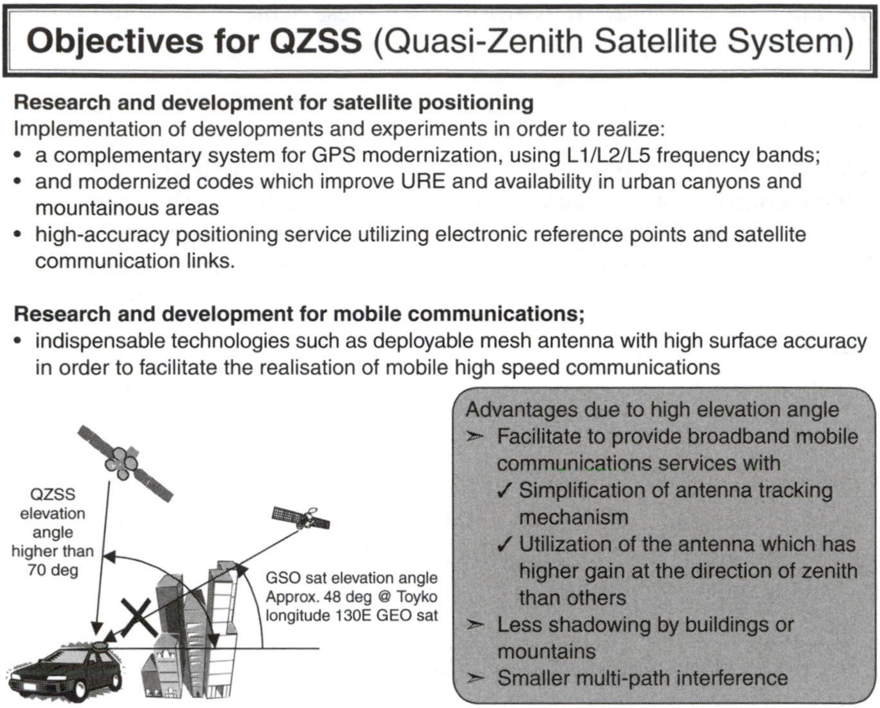

Figure 4.3 Objectives for Quasi-Zenith Satellite System (QZSS) (source: Japan GPS Council).

creation of new demands for updated replacements. In FY 2001, new car navigation products with hard disk drive (HDD) were widely accepted, and reached 2.19 million units, or a 15 per cent increase from the performance in the preceding year. In FY 2003, the yearly shipment of car navigation products exceeded three million units.

The FM-DGPS loading ratio had shown a steady growth even after the turning off of selective availability (SA). Car navigation seems to be the fountain of synergistic effects, as represented by sophisticated products for new and trendy 'car multimedia' that combine DVD or HDD display, mobile communications, traffic control information and other attractive contents. New mobile multimedia markets consisting of 'Location + DVD/HDD Theater + Telematics' are emerging rapidly in Japan, as the result of combined car navigation with VICS services. VICS, one of the important elements of Intelligent Transport Systems (ITS) in Japan, is a public service that transmits traffic congestion information and route guidance to the car navigation display via three media: multiplexed FM, wireless and optical beacon links. The VICS centre of the service provider is expanding the service area in anticipation of enhanced demand in local cities.[9]

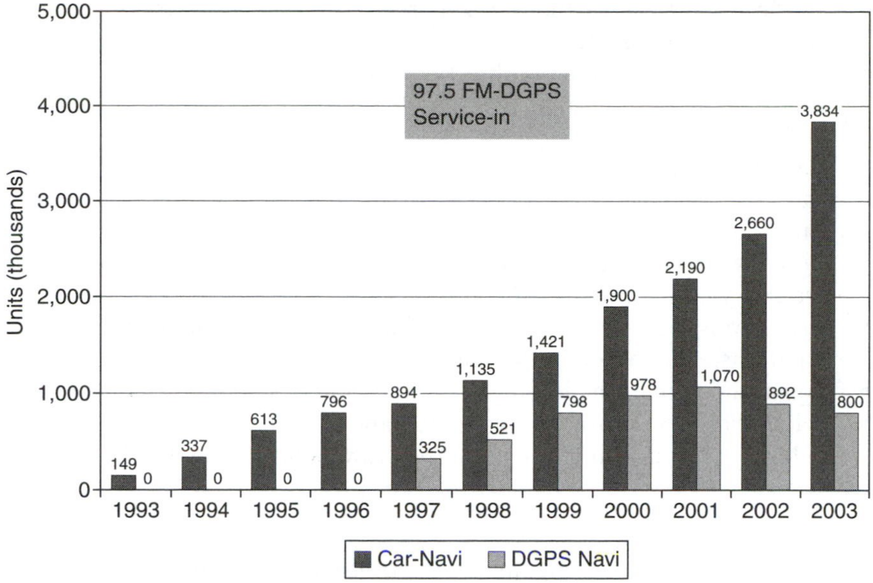

Figure 4.4 Car-navi market in Japan, 1993/94–2004/04 (source: Japan GPS Council).

Note
The figure for 2003 shows data from 2003.1 to 2003.12 and includes OEM export.

Emergency call service, HELPNET

The emergency call service HELPNET (Help systems for Emergency Life saving and Public safety Network), similar to the United States' E-911 was launched in mid-2000. The service provider, Japan Mayday Service Co Ltd, was established, after a year-long study conducted by the National Police Agency and Ambulance Agencies, as a private-sector entity with investment from 38 major companies, including car manufacturers, electronics, mobile phone carriers, insurance, banks and security guard entities. The business of the company is to provide subscribers (mainly car drivers) with emergency information services by connecting them to an adjacent police or ambulance office. The service is provided with GPS/DGPS-based accurate location data using cellular devices such as in-car mobile phones.

Great impact of SA being turned off continues

New GPS-related business launchings have been reported in newspapers almost every day since May 2000. The driving force is the improved GPS positioning accuracy and stability of signals without SA and the expansion of wireless mobile phone markets. This has led to the rapid expansion of new businesses for:

- ASP for car navigation and man navigation;
- traffic hazard forecasting (frozen roads/area weather, traffic congestion, hazard maps, etc.);
- thunder observation network service;
- kinds of service used in the search for roaming elderly people, or stolen vehicles;
- fleet management;
- bus fleet management services (on-demand buses, efficient shuttle buses, etc.);
- new taxi services (tour guides, shopping guides, transportation to and from day-care centres, etc.);
- monitoring of waste collection vehicles (chemicals/industrial waste/solid waste, discarded television sets, etc.);
- construction machine leasing network with GPS telematics;
- telematics services for car navigation and mobile phone users;
- robotic construction (full-automatic crawler drill/IT hydraulic excavator, etc.);

Location-based services using handheld positioning and navigation capability

The 3G-model mobile phone au™ with incorporated GPS capability, put on the market at the end of 2001 by KDDI, one of the mobile phone service providers in Japan, has gained a wide acceptance, and the number of subscribers already exceeded eight million by November 2003. (See Chapter 3 for further elaboration. The security service providers that utilize au™ are also gaining subscribers rapidly. HELPNET, mentioned earlier, is expanding the range of target customers for au™ in addition to the car navigation users. Many entrepreneurs are emerging.

National point of contact: for civil associations

The Japan Global Positioning System Council has been playing an important role in assisting Japanese policymakers in relevant fields. It has been, and expects to be, one of the valuable information resources for GPS and telecommunications.

Conclusion

In conclusion, the overriding influence in the development of GPS and satellite in Japan is not just technical but geopolitical in nature. The Japanese satellite positioning system is terrestrial based rather than based on satellite navigation like GGPS. QZSS cannot compete with GGPS and the American GPS system unless it is profoundly improved, and the Japanese satellite programme, to which it is linked, will not be able to compete with

Galileo either. This leaves a possibly delicate situation if GPS and Galileo do not cooperate and remain competitors. Japan, India and Brazil may go with an inferior system, GPS, while China, Europe and possibly the Middle East will go with Galileo GPS. Or the giants China and India could develop their own shared system, unbalancing the world's information-sharing structure. This could have major ramifications for power balances, trade, and levels of technical expertise around the world. Japan will be placed in a very difficult situation, being forced perhaps to align itself with the Asian camp or with the Western nations (which it has preferred to do to date). As China and India develop increasingly close ties, it is most likely that Japan and the United States will not be welcome members of their joint projects. Tension has been mounting steadily between Japan and China for the past year over intellectual property infringement, North Korea and other issues, culminating in April 2005 in anti-Japanese riots in major cities of China and South Korea over a textbook distorting the destructive role of Japan in China and Korea during the Second World War. China could also use political leverage gained against Japan and in alignment with India to consolidate its hold on Tibet and take over Taiwan with impunity.

However, it remains to be seen whether all these efforts can be coordinated to offer the world the best possible GPS services, bringing Japan, the United States and Europe closer together, but not in an exclusionist manner that would deprive already disadvantaged countries of much-needed information sources, or pushing the advanced countries further apart, which would also be counterproductive from many perspectives. One outcome is clear, independent of all the possible permutations in alignment, and that is that whoever controls the most advanced satellite/GPS telecommunications information networks will have an inordinate amount of power globally.

Notes

1 Dr Naokazu Hamamoto, Research Supervisor, Wireless Communications Division, CRL, 'Activities for Advancement of Satellite Communication at CRL', Fifth International Forum on Advanced Satellite Communications in the Asia Pacific Region Report. November 2003). Some of this material is also from the *Soumusho* (MIC), where I conducted a number of interviews in April 2004.

2 Suguru Hattori, 'Signal Success' *Look Japan*, 49 (575) (February): 26–28 (2004).

3 Some of this material derives from discussions with the *Soumusho* (MIC) in early April 2004: Richard Garwin's article in *Foreign Affairs*, entitled 'Galileo and GPS – Cooperation or Confrontation?' (November 2002, Japanese edition). Much of the material concerning the US attitude to GPS derives from material sent to me by F. Michael Swiek, Executive Director, United States GPS Industry Council.

4 Information source: Terry Moore, Professor of Satellite Navigation and Director, Institute of Engineering Surveying and Space Geodesy, University of Nottingham.

5 The updated national report from Japan is by Hiroshi Nishiguchi, Secretary-General, Japan GPS Council, 44th CGSIC, 20–21 September 2004 at Long Beach, California.
6 CGGTTS: CCDS (the Consultative Committee for the Definition of the Second) Group for GPS Time Transfer Standardization.
7 It is clear why Japan was obliged to change its geodetic system standard; see below http://mekira.gsi.go.jp/English/crstanime.html).
8 Vehicle Information and Communication System: traffic jam information and route guidance services, one of the ITS programmes.
9 According to the VICS Centre URL, the number of unit shipments in 2003 was 2.4 million and the cumulative total reached 8.46 million.

5 Spectrum policy

Hajime Oniki

1 Introduction and background

This chapter gives an overview of spectrum policy in Japan. It is written mainly for readers who are interested in this subject but have little information on it other than from the general media or from IT periodicals.

There are four sections in this chapter. A brief introduction to recent issues in spectrum policy in general (i.e. not limited to that in Japan) is given in the remainder of this section. The second section summarizes the present state of spectrum regulation in Japan, with some historical background. The third section is the main body of this chapter; it is devoted to explaining recent spectrum policies in Japan from the late 1990s to date. This section is composed of five subsections, each covering a specific aspect of spectrum policies. There are two appendices to this chapter. Appendix A gives a list of revisions made to the Radio Law of Japan, which will serve as a concise summary of recent spectrum policies. Appendix B gives a chronology of major events in spectrum regulation in Japan, which may be examined along with the third if the reader is interested in its historical aspect. In the final section, attempt to explain why in Japan, in contrast to other countries, spectrum policies are still conducted under what is called command and control, not by market mechanisms such as auction. This section is sketchy and in no sense complete; it is written mainly as a basis for further research on this subject.

1.1 Spectrum as a resource

This and the following subsections give a brief outline of the background of issues in spectrum policy. Those readers who are familiar with recent events such as spectrum auction and trading in the United States, Europe and other countries may skip this section. Those who wish to study more on this subject may begin with, for example, the first chapter of Milgrom's recent book[1] or a report published by the Federal Communications Commission (FCC) of the United States.[2]

To begin with, we note that the use of radio spectrum has been under the control of central government in almost all countries. Spectrum policy in advanced countries became important in the late 1980s and the early 1990s, when the demand for spectrum increased rapidly because of the growth of mobile telephony. Before that time, new spectrum was supplied successively as demand arose; spectrum shortage was rare. Today, however, it is a major policy objective to satisfy ever-increasing demand for spectrum.

Radio spectrum, considered as an economic resource, is one of the space resources, of which other examples are land space, water-surface space and airspace. The value of land (i.e. a portion of the surface of the globe) as an economic resource arises from its being used physically for various activities such as agriculture, housing, factory production and transport. The value of terrestrial spectrum (radio spectrum used on the surface of the globe) as an economic resource arises from its being used electromagnetically for communication, radar, and other purposes.

For this and other reasons, the economic properties of spectrum resemble those of land; spectrum can be treated much like real estate. Thus, we can consider an 'area' of spectrum with its 'boundary', we can divide spectrum into smaller pieces, and we can consider the 'capacity' of an area (called a band or block) of spectrum. Further, it may be decided to establish rights in connection with spectrum so that it can be bought, sold or leased at a price. Of course, spectrum can be given away free of charge, or can be shared by many users.

A major difference between land and spectrum lies in that the use of land is limited usually to a single objective, whereas there can be many spectrum uses on an area of the global surface, since we can use alternative radio frequencies for different purposes. (For example, we can view multiple television channels while using a mobile phone at one location.) One way to state this is to say that there is only one physical space on an area of the globe, whereas there are many spectrum spaces on the same area.

1.2 History

Radio spectrum was first used about 100 years ago for navigational safety and naval operations. Since that time, the utilization of spectrum has expanded steadily and greatly. Voice radio became popular in the 1920s, and radar was invented in 1935 during the lead-up to the Second World War. In the 1950s, television receivers, first black-and-white and then colour, became a major household good. Today, in many countries, mobile telephony shows penetration exceeding half of the population and spectrum is used widely for many other purposes.

Such remarkable development of the utilization of radio spectrum was achieved, needless to say, by a succession of technological advances.

Typically, a new technology was introduced by making use of a new 'band' of radio frequencies which had so far been unused; that is to say, the development process was an expansion of the frontier of spectrum utilization.

During the course of this process, the use of spectrum was under direct control by central government in almost every country. For one thing, the major concern was how to prevent interference between spectrum users; this called for public regulation. For another, spectrum was first used for safety and security; it was natural for the government to play the role of managing it. Furthermore, since new utilization of spectrum was made possible by the expansion of spectrum frontier, spectrum scarcity was not a major concern; the government was able to award the right to use spectrum without any trouble. Thus, until recently the principle of command and control (C/C) by the government prevailed in spectrum utilization.

This situation, however, has now changed. As the speed of technological progress increased, the demand for spectrum grew exponentially in view of new services such as mobile telephony and wireless Internet access. Roughly speaking, at the beginning of the twenty-first century, the frontier of economically usable spectrum was nearly exhausted. The present situation is such that we are unable to find a frequency band for new services in the same way as we were able to in the past.

It should be noted, however, that, while the frontier may have been exhausted, this does not mean that there is no way to find additional spectrum for new services, a large proportion of the spectrum bands remains unused, or used very inefficiently. During the time of frontier expansion, it was of little concern for the government to have spectrum used efficiently. Furthermore, it was not a concern of users either to save spectrum, since the price was near zero. As a consequence, the state of utilization of radio spectrum at the present time is in extreme disequilibrium; some spectrum is used efficiently with a large amount of expenditure on new equipment, but other spectrum is used inefficiently with old and obsolete equipment. This situation may be compared to one in which large areas of farmland were to be found within the City of London or next to the Empire State Building in New York. Such a case in the use of land would be precluded by market forces. For radio spectrum, we do have extremely unbalanced utilization, since market forces are not working.

1.3 Systems for using spectrum

Currently, in almost all countries radio spectrum is brought to use in two stages: spectrum allocation and assignment (licensing), each controlled by the government. In the process of spectrum allocation, the entire spectrum of radio frequencies is divided into spectrum bands, for each of which an objective of use is specified.[3] Spectrum allocation is conducted with international coordination through the World Radio Communication

Conference (WRC). In the process of spectrum assignment, which is conducted domestically, a spectrum band is further divided into a number of spectrum blocks, frequency-wise and area-wise, to each of which is assigned a user with a licence issued by the government.[4] In comparison with the use of land, we may state that spectrum allocation corresponds to 'zoning' in city planning, whereas spectrum assignment corresponds to establishing the right to use a piece of land.

The two-stage system, allocation and assignment, was developed through C/C at a time when there was no spectrum shortage. It has become clear, however, that C/C cannot function smoothly now that there is spectrum shortage, since it is difficult and time-consuming to deal with competing demands for spectrum.

Attempts to find a new system for spectrum utilization started in the 1980s and still continue today. Most such attempts have been made within the framework of the two-stage system, rather than by abolishing it in favour of something entirely new.[5] That is to say, new systems have been offered and used by changing the actual functioning of spectrum allocation or assignment, not by changing the formal system. Further, most of the new systems that have been offered are concerned with spectrum assignment, not allocation.[6]

Traditional C/C means that the government retains the authority to select the user of a spectrum block. When there is only one applicant, the government grants a licence as long as the applicant satisfies certain prescribed qualifications. When there are two or more applicants, the government selects the one (from among those qualified) that can best serve the people by using the spectrum block (comparative hearings).

In 1981, the US Congress voted for the introduction of random selection (lottery) to replace C/C in order to assign spectrum for first-generation mobile telephony (analogue mobiles). This system attracted some hundreds of thousands of applicants for a licence. It is reported, moreover, that the system reduced the time needed for assignment to two years from the four to five years taking under C/C.

The first auction for spectrum assignment was conducted in 1990 by the government of New Zealand. The United States followed in 1994 with a series of auctions beginning with the ones for the second-generation digital mobiles (PCS). Since then, the FCC has conducted more than 50 auctions to assign spectrum for various uses; most of them have been successful, but there were a few major failures. In the years 2000 and 2001, the United Kingdom, Germany and other European countries assigned spectrum for third-generation mobiles (3G) by auction; in the United Kingdom and Germany, however, successful bids for licences were extremely costly. No conclusion has yet been reached on whether this was a successful outcome or not. Meanwhile, countries including Australia, Brazil, Singapore and Hong Kong also conducted spectrum auctions for 3G.

Today, we are in the process of searching for a useful system of spectrum allocation and assignment (or rather, spectrum reallocation and reassignment). One such system may introduce spectrum trading or spectrum lease. These systems, together with auction, rely on market transactions of spectrum with price – what is called the 'market mechanism'. In an extreme case, a block of spectrum may be treated as private property in much the same as a piece of land is; this system is simply called 'property'.[7] Introduction of the market mechanism is supported by many scholars and experts today because of its capability to realize the efficient use of spectrum resources.

In contrast to this, there is a group of specialists who advocate the introduction of 'spectrum commons', in which a spectrum block is offered to the public for free use (but within some prescribed rules or 'etiquettes').[8] Spectrum commons may be compared to city or town commons or public parks as a way to share an area of land for open use. While commons use of spectrum may bring many advantages, it has a major drawback: the possibility of congestion and interference ('the tragedy of the commons'). Roughly speaking, commons use is supported by engineers who believe in the capability of spectrum-sharing technology, including that in the future, to accommodate as many users as desired within a given block of spectrum, however small. Some of them even claim that spectrum regulation is no longer needed because of such technological progress.[9] A majority view, however, is that while spectrum sharing with commons may work in cases in which spectrum is used with weak radio emissions (as in 'unlicensed band' today), some non-shared use (exclusive use) of spectrum is required when strong radio emissions are used (such as in broadcasting and radar), and that if that is so, we need a market or, at the very minimum, C/C for spectrum assignment.

Thus, at the present time there are three candidate systems for spectrum assignment: the traditional C/C, market and commons. Although no agreement has been reached widely, the majority view is that C/C cannot function well once spectrum shortage emerges; the selection of market or commons should be made in consideration of the service to be provided with spectrum and underlying technology.

2 Current regulation of spectrum in Japan[10]

2.1 Outline

The use of radio spectrum in Japan is regulated by the Ministry of Internal Affairs and Communication (MIC) under the Radio Law. The overall design of regulation is the same as in other countries; radio spectrum is brought to use in two stages: spectrum allocation and assignment (licensing), each controlled by the of MIC. MIC still uses C/C for spectrum assignment. In Japan, however, there have been a few cases in which a spectrum licence was sought competitively by two or more applicants. It is understood that in such a case, MIC performs some informal adjustments beforehand so that there is only one applicant at the time the assignment decision is made formally (see p. 124). (MIC officers would deny this, though.)

Since the late 1990s, MIC has been officially maintaining that 'spectrum assignment by using a market mechanism such as auction is not desirable because of the possibility of excessively high bids (as in some of the 3G auctions in Europe), which may delay the deployment of a new service'. Incumbent spectrum users in Japan would agree with this, whereas many experts and academics would suggest that MIC should use auction for spectrum assignment. As I have stated, in the United States, the United Kingdom and other countries, spectrum assignment is often conducted by means of a market mechanism such as an auction; the MIC's determination to continue using C/C is unusual among advanced countries. Since, however, the demand for spectrum is rapidly increasing these days because of the progress of spectrum-using technology, MIC may be forced to change its policy for spectrum assignment in the near future (see pp. 124–125).

2.2 Using spectrum

The user of spectrum in Japan needs to secure a licence from MIC to establish a wireless station. A licence specifies, among other things, the name of a user, a block of spectrum to be used, the objective of using the block (i.e. the band to which the block belongs), the location of the wireless station, the limit of power emission with other technical constraints, and the duration of the licence (licence term), which is usually five years. It is expected that a licence can and will be renewed upon expiration unless something extraordinary takes place.[11] The licensee may be an individual person or a corporate body. A non-Japanese licensees are not allowed.[12] Further, a licence cannot be sold or given to another party except when a corporate body holding a licence is involved with mergers and acquisitions (M&A; see p. 122). This effectively excludes the possibility of (secondary) licence markets in which a licence can be traded freely without the permission of MIC.

There are a few exceptions to the licensing system as summarized in the previous paragraph. First, when a licence is used for the safety of navigation or aviation according to an international treaty, the licence term is indefinite and there is no need for renewal. Second, no licence is required when the power emission is very low or when using certain designated spectrum bands (unlicensed bands). Third, some spectrum bands may be used with only registration at MIC (i.e. without a licence) (see p. 123). Fourth, a branch of the Japanese government may use spectrum without a licence with the agreement of MIC. In particular, the use of spectrum for military purposes, including the use by US military forces stationed in

Japan, is outside the scope of the Radio Law.[13] Prefectural and local governments as well as other public agencies, however, need a licence to use spectrum, although they may be entitled to special treatment depending on the objective of spectrum use.

Currently in Japan, mobile telephony and broadcasting are the two major objectives of spectrum use. Approximately two out of every three Japanese carry a mobile phone, and two to eight terrestrial television channels can be viewed almost everywhere in Japan. The total number of licence holders, excluding those for mobile telephony, is approximately 2,000,000.

2.3 Spectrum regulation

The legal basis of spectrum regulation in Japan is the Radio Law, supplemented by a number of administrative and operational rules such as ministerial ordinances. The Radio Law gives MIC the authority to regulate spectrum use, and thus MIC plays the central role in forming and executing spectrum policies in Japan.

MIC is formally headed by the Minister of Internal Affairs and Communication and a few vice-ministers, all of whom are political appointees. The rest of MIC is an organization of civil servants. MIC covers a large variety of public services and regulations, including those in wireline and wireless communication. Figure 5.1 is an organization chart of the communications part of MIC; it has two bureaux, the Information and Communications Policy Bureau and the Telecommunications Bureau. The Radio Department in the latter conducts spectrum regulation. The Radio Department has five divisions: the Radio Policy Division, the Fixed Radio Communications Division, the Mobile Communications Division, the Satellite Mobile Communications Division and the Radio Environment Division. The Radio Policy Division is central in planning spectrum policies; the Usage-Fee Planning Room, one of its two subdivisions, collects spectrum user fees and also maintains a database of spectrum licences. The next three divisions in Figure 5.1, the Fixed Radio Communications Division, the Mobile Communications Division, and the Satellite Mobile Communications Division administer the use of spectrum for fixed terrestrial, mobile terrestrial and satellite communications (including all communications for marine navigation and aviation), respectively. Finally, the Radio Environment Division is responsible for maintaining the safety in using the radio spectrum. Its subdivision, the Radio-Usage Monitoring and Enforcement Room, protects spectrum users from radio interference and other improper use of spectrum.

The broadcast divisions in the Information and Communications Policy Bureau, including the Broadcasting Policy Division, regulate broadcasters; they cooperate with the Radio Department for spectrum-related matters in broadcasting.[14]

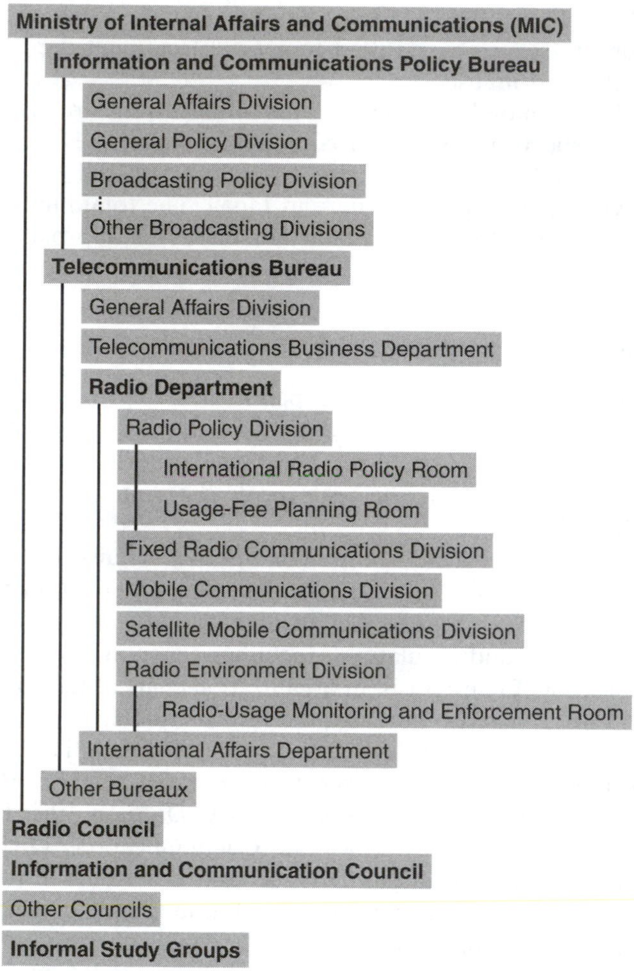

Figure 5.1 Organization chart of the Ministry of Internal Affairs and Communications radio administration (source: http://www.soumu.go.jp/joho_tsusin/eng/aboutus.html, translated and modified by the author).

The Radio Council and the Information and Communications Council are organizations formally outside MIC; the two councils, however, work in close relation with the Ministry. A council is composed of a small number of experts nominated by Parliament on the recommendation of MIC. The Radio Law requires the MIC to consult with the Radio Council on certain important issues in spectrum regulation. The Radio Council also handles complaints and lawsuits between spectrum users and between a user and MIC.

In addition to the Councils, informal study groups, initiated by MIC, are frequently in action. When MIC is considering forming or reforming a spectrum policy, it creates one or more study group(s), of which the composing members are experts (such as university professors) and interest holders (such as representatives of spectrum-using companies); they are nominated by MIC. MIC uses a study group to collect technical and other information useful for spectrum policy formation, to reconcile confronting demands from interest holders, to explain and justify policies proposed by MIC to the general public, to collect comments from the general public, and to form an outline of draft bills for a new policy. Meetings of a study group are held in an informal manner. In most cases, officers from MIC serve as the executive secretariat of the study group and lead the discussions in it, often through controlling the chair.

The following is a typical procedure by which MIC implements a major policy for spectrum regulation (for example, starting a new service such as IMT-2000).[15]

1. MIC conducts some investigation of the matter internally and forms a preliminary plan for a new policy. At this point, the baseline of the new policy is fixed by MIC; it is extraordinary for MIC to change it after the process is started, although minor adjustments may be made later, depending on the response from study-group members or from the public. (This, however, is an understanding by those outside of MIC; MIC officers would not comment even on the existence of such an internal process.)
2. An informal study group is formed to consider the matter.
3. The study group meets, say, once or twice a month for, say, six to 12 months. It is customary to disclose (on the Internet) at least an outline of the minutes of each meeting, but often without including the names of speakers. Materials distributed at the meeting are published (on the Internet), too. In effect, anybody who so wishes can trace the activities of a study group in detail.
4. A draft of the final (or interim) report, which outlines the new policy, is written in the study group. Usually, MIC controls drafting a report through its secretariat. (Usually the secretariat, not study-group members, writes a draft.)
5. The draft report is published for comments from the general public. The period for submitting comments is one month or so.
6. Public comments are assembled and disclosed. At the same time, MIC publishes a set of ‘MIC replies and explanations’ to an edited outline of the comments collected. Often, the comments are abridged and edited so as not to block the baseline of the new policy as set out by MIC initially. Reply comments are seldom invited.
7. A final (or interim) report is assembled. Usually, public comments are taken into consideration with regard to minor points, but not to the

major content of the prescribed policy. The study group is closed at this point.

8 If required by the Radio Law, the content of the report is discussed with the Radio Council; usually, it is approved smoothly by the Council as a recommendation.

9 MIC drafts a legislative bill to amend the Radio Law based on the recommendation by the Radio Council. (Note that this is the first time that the new policy is written in the form of a legislative bill; the activities of a study group and the Radio Council are conducted on policy outlines only.)

10 MIC proposes the bill to the cabinet, headed by the Prime Minister; approval is automatic in effect.

11 The bill is presented to Parliament. Bills amending the Radio Law are almost always approved by both houses of Parliament (as a consequence of the control of the houses by the Liberal Democratic Party (LDP)).

12 The bill becomes a public law, and the policy initiated by MIC is implemented. On average, the entire procedure takes one to three years.

It is customary that when MIC makes a rule (such as an MIC ordinance) or a plan (such as a spectrum assignment plan), the process follows a path similar to that just described, although it is possible for part of the process to be omitted if the matter is not important.

All in all, MIC's activities in spectrum regulation (and other matters) are conducted fairly transparently today, although tracing even a portion of the regulation is costly and time-consuming, since the system is complicated and not yet well organized.

2.4 The Radio Law

The Radio Law is legislated, needless to say, by Parliament (the Diet). In Japan, however, most legislative bills are drafted in the executive branch of the government. In the case of the Radio Law, its initial legislation and almost all the amendments hitherto have been written, and proposed to Parliament, by MIC; further, all the bills amending the Radio Law have been approved by the Parliament without modification. In effect, MIC has not only executive but also legislative powers on spectrum regulation.

The Radio Law as amended last in 2004 has 11 chapters altogether: Chapter I: General Provisions; II: Licenses for Radio Stations; III: Radio Equipment; III-2: Technical Regulations Conformity Certification, etc. of Specified Radio Equipment; IV: Radio Operators; V: Operations; VI: Supervision; VII: Protests and Lawsuits; VII-2: Radio Regulatory Council; VIII: Miscellaneous Provisions; and IX: Penal Provisions. Articles determining spectrum policies are to be found mainly in Chapters I, II, VII, VII-2 and VIII, among which Chapter VII-2 specifies the organization and the function of the Radio Council, and Chapter VII establishes the procedure for handling protests and lawsuits by the Radio Council. The remaining chapters are for technical and operational regulations.

It was in 1950 that Radio Law was first enacted; Japan was still under the occupation of the Allied forces of the Second World War led by the United States, which ruled Japan indirectly through the Japanese government. Accordingly, the Radio Law was strongly influenced by US communications law. Thus, the main regulator of spectrum in the original 1950s Radio Law was the Radio Council, functioning in an open-court style; it was not a ministry, which would have prevented the process of internal decision-making from being disclosed. In 1952, soon after the Radio Law was first enacted and shortly before the Allied occupation was ended, the authority for spectrum regulation was transferred to a ministry (which became MIC), except that the function of handling protests and lawsuits was left with the Council.[16] Thus, Chapters VII and VII-2 today, which specify how the Radio Council handles protests and lawsuits, are descendants of the system of 1950. For this reason, the Radio Council, together with the Fair Trade Commission and a small number of other organizations, is one of the best-legislated court-style organizations within the executive branch of the Japanese government. However, the ministry system is still solid in Japan today, and a majority of MIC's actions are conducted under the traditional ministry system, not with a court-style system.[17]

For about three decades after the Radio Law was first enacted, the main focus of spectrum regulation in Japan (as in other countries) was on the technical and operational side. The wireless business grew steadily but slowly. There was a good amount of spectrum supply as a consequence of technological development, which released new spectrum bands such as those in VHF and UHF one after another.

It was the explosive growth of mobile telephony that changed the state of spectrum resources from excess supply to excess demand. In Japan, the rapid increase in mobile subscribers began in the second half of the 1990s, a few years later than in European countries and the United States. It was followed by additional demand for spectrum in the late 1990s and early twenty-first century for digital television (DTV), local area networks (LAN), radio frequency identification (RFID), and countless new wireless services, all brought in by the development of digital wireless technology. As is explained in the following section, MIC's spectrum regulation has become busier since the mid-1990s.

Figure 5.2 shows the change in the 'size' of the Radio Law, as measured by the number of (Chinese and Japanese) characters used in the text of the law, for the period 1950–2004. Since a new spectrum regulation accompanies the addition of new articles to the Radio Law, its size is a good, if not precise, measure of the degree of regulatory activities:

We can distinguish two periods in Figure 5.2. During the first, 1950–1985, the Radio Law grew slowly. Most of the increase during this

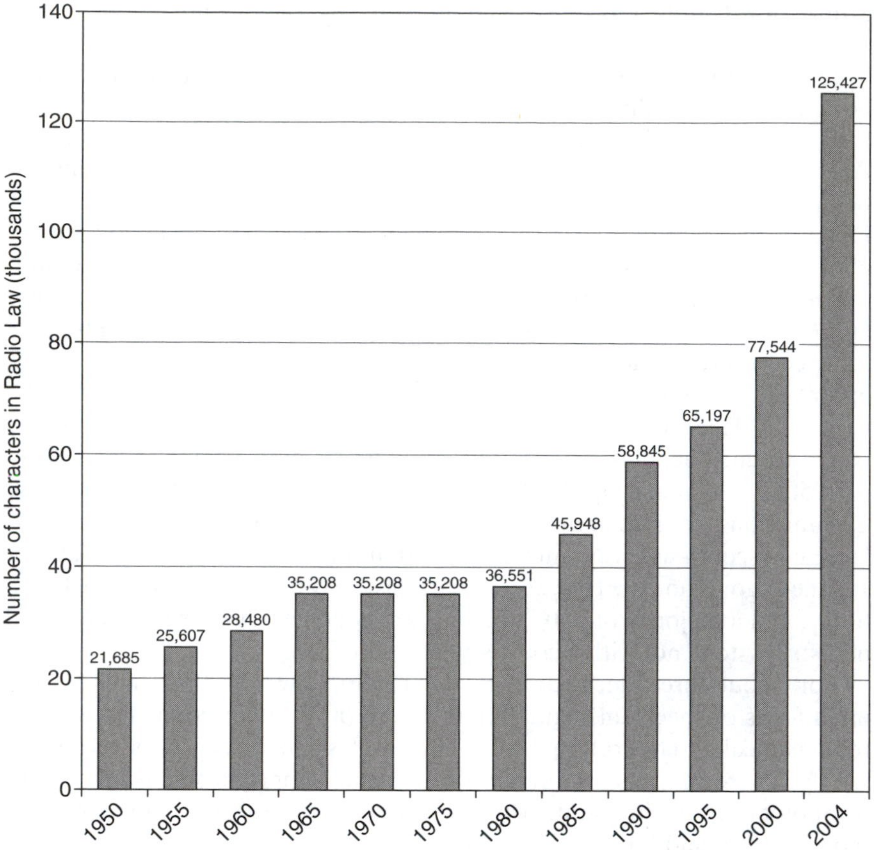

Figure 5.2 'Size' of the Radio Law, 1950–2004 (source: compiled by the author based on information from data in web pages of the Japanese Parliament, http://www.shugiin.go.jp/index.nsf/html/index–housei.htm).

Notes

1 Size of law each year is represented by the number of characters included in the law.

2 The size of law in 1950 and that in 2004 are actual numbers. The size in other years is estimated by the author with allotting the size of each of the twenty-five laws which amended Radio Law for the period between 1950 and 2004.

period was directed to the expansion and improvement of technical and operational regulations. By contrast, the increase during the second period, 1985–2004, is rapid and accelerating. More than half of the increase during this period is for administrative and economic regulations to deal with the increased demand for spectrum. A curve fitted to the graph of Figure 5.2 might suggest a formidable increase in the administrative cost (transactions cost) incurred from the micro-management by MIC with C/C for spectrum regulation in an age of the rapid growth of wireless services.

Recent spectrum policies

Outline

This section will overview major spectrum policies conducted by MIC during recent years. The section is arranged by subject, not chronological order. Appendix B to this chapter, 'Chronology: Japanese spectrum policies, 1992–2004', lists major events in spectrum regulation and others in chronological order with cross-references to section numbers in the main text of this chapter.

As is stated in the preceding section, MIC has been using C/C to allocate spectrum bands and to assign spectrum blocks to users. It may become difficult for MIC to continue this in the future. First, selection of a licensee by means of C/C may be challenged through repeated appeals (litigation) beyond a limit that can be handled by MIC. (In 2004, we may be entering such a stage; see p. 125). Second, the cost of regulation (transactions cost) may become prohibitively high. In order to cope with the increased demand for spectrum, MIC has several times revised the Radio Law, together with MIC's regulation rules, which is what has led the Radio Law's ever-increasing size. MIC will then be forced to revise the system at some time in the future because of the increased complexities of regulation.[18]

However, MIC has so far been doing quite a good job of regulating spectrum within the C/C framework. Amendments to the Radio Law have been meticulously detailed to handle various situations and cases, and with great care being taken to maintain consistency within itself and with other laws.[19] MIC has been able to respond to new demands for spectrum so as to keep the overall efficiency and equity of spectrum utilization within the limit of C/C.

The following subsection summarizes what MIC did during the period 1992–2004, by grouping various events and actions concerned with spectrum regulation into four categories: (1) adjusting the regulation system for increased use of spectrum; (2) preparing spectrum for new services such as IMT-2000, DTV and wireless LAN; (3) planning overall reallocation of spectrum; and (4) charging spectrum user fees.

3.2 Adjusting regulation for increased use of spectrum

3.2.1 Introduction of blanket licences, 1997, 2000

In 1997, soon after mobile telephony began expanding with an annual increase of a millions of subscribers, MIC introduced a system called a 'blanket licence' to simplify its licensing procedure (RL27-2 to 27-11).

In Japan, a spectrum licence is formally considered to allow the establishing of a wireless station emanating radio waves, rather than the use of radio waves with designated frequencies, areas, time of use, etc. Thus, in principle, anybody who operates a device emanating radio waves needs to apply for a licence, including mobile telephone subscribers. This, however, is too costly, because of the tens of millions of subscribers.

With the blanket licensing system of 1997, MIC can grant a (blanket) licence to mobile operators to establish base stations and to sell mobile terminals with the understanding that all subscribers are excused from obtaining individual licences. This has greatly simplified the licensing procedure and saved costs for both MIC and operators.

In 2000, MIC further simplified the licensing for mobile operators with the introduction of microcells (RL27-2 to 27-17). Microcelling is a way for an operator to accommodate additional subscribers in a region with a given size of spectrum blocks. A cell is a region covered by a single base station (a single antenna). Microcelling divides a cell into several cells of smaller size (microcells). MIC's new system for base stations (i.e. new micro stations) eliminates the need for an operator to obtain a licence for each micro station; instead, the operator can obtain a blanket-type licence (obtain permission to establish specified base stations) to create microcells.

There can be no objection to the introduction of the blanket licensing schemes of 1997 and 2000, since these saved costs for both MIC and mobile operators. However, in a sense these simplifications have eroded the principle that MIC firmly regulates the use of radio waves by issuing a licence for the construction of a wireless station (i.e. by regulating the behaviour of a spectrum user directly); the arrangements have introduced the idea that an operator has the right to use a spectrum block exclusively with freedom to determine the details by itself (rather like using real estate). That is to say, the arrangements brought in the idea of considering spectrum as property, for which user rights could be prescribed.

3.2.2 Licence transfer in the case of mergers and acquisitions (2000)

Originally, the Radio Law of 1950 allowed a corporation to transfer, or succeed to, a spectrum licence only when that corporation merged with another corporation. In the 1980s and 1990s, however, mergers and acquisitions (M&A) of corporations became popular in Japan; in view of this trend, MIC deregulated its restriction on licence transfer. When a corporate is divided into more than one corporation, one of the new successor corporations can take over the spectrum licence with the permission of MIC (RL20(2) (3)).

As I have said, from the standpoint of administrative and economic efficiency, there is nothing to object to in this deregulation. However, it has introduced the idea that a licence represents the right to use spectrum, and that this right can be 'sold' as a property, since the 'price' of a licence would be included in the value of the new corporation at the time of M&A. This is a backdoor infiltration of market mechanisms into C/C. Whereas the original licensee obtained a licence without paying its economic value (a licence is granted free of charge by MIC), its successor body may insist that the licence was purchased, in that a price for it was included in its valuation of the new corporation. This goes against the idea of the Radio Law that a spectrum licence is a privilege awarded to a person, not a property to be purchased at a price.

3.2.3 Wireless station established by registration, 2004

A new system for establishing a wireless station by means of registration was created by MIC in 2004 (RL27-18 to 27-34). This is concerned with the use of spectrum for providing wireless access to the Internet (wireless LAN). When the emission power is very low, as in wireless access within a building or with a cordless telephone, unlicensed bands have been used. For outdoor wireless access, however, some degree of power emission is needed and unlicensed bands cannot be used. Further, spectrum is to be shared with other providers – hence the new system, under which providers can open wireless stations for outdoor Internet access only by registering with MIC.

In addition, the Radio Law was amended at this time to make it possible for outdoor wireless LAN subscribers to open their terminal under a blanket registration. The relation between a provider station for outdoor wireless LAN constructed by registration to an LAN user's wireless terminal is the same as the relation between the base station of a mobile telephone operator and a terminal held by a mobile telephone subscriber. Thus, under this blanket registration system, wireless LAN subscribers need not register their use of wireless LAN devices individually.

Congestion and interference among wireless LAN operators and subscribers might arise under this system if too many operators attempted to open base stations within a narrow area or if too many wireless LAN subscribers attempted to use LAN within an area. Such is one of the shortcomings of spectrum sharing, the tragedy of the commons. For unlicensed bands, no measures are taken to prevent congestion or interference. For spectrum used with registration, the Radio Law gives MIC the authority to block new registrations if it expects congestion and interference in a particular area (RL76-2-2). Thus, the use of spectrum with registration as introduced in 2004 has characteristics both and of exclusive use.

3.3 Preparing spectrum for new services

3.3.1 IMT-2000 (3G mobile phones)

a. OUTLINE

IMT-2000, third-generation mobile telephony (3G), is a mobile service with capabilities far greater than that of second-generation mobile telephony, which is the service most used worldwide in 2004. MIC has been eager to prepare a framework, spectrum and others, for the introduction of 3G mobiles. The WRC allocated spectrum bands to be used for IMT-2000 by its member countries in the late 1990s. A set of technical standards for 3G was developed in the International Telecommunication Union (ITU) through the second half of the 1990s. The ITU, however, was not able to recommend a single standard for 3G; it was reported in 1997 that at least two code division multiple access (CDMA) standards would be used for the 3G service. In 2000, the Ministry announced that there would be three spectrum blocks to be blanket-licensed nationwide for IMT-2000, corresponding to the three major 2G operators in Japan.

b. THE CASE OF QUALCOMM, 2000

In May 2000, a Japanese subsidiary of the US company Qualcomm, Inc. announced its intention to apply for a 3G licence with MIC, soon after the three major 2G operators disclosed that all of them were planning to use the CDMA standard, which was supported by Japan and European countries in the ITU (W-CDMA). Qualcomm, the owner of the CDMA 2000 technology, which the United States supported in the ITU as a 3G standard, would have no market in Japan if CDMA 2000 were used by none of the three operators. Thus, at this time there appeared to be four applicants for the three 3G licences. This was probably the first time that the Ministry had faced an apparently competitive situation in assigning major spectrum blocks.

It was understood that the Ministry was firmly determined to use C/C, not a market mechanism such as auction, for the 3G market. It was inconceivable that one of the three Japanese operators would be excluded from the 3G market. Yet it was difficult for the Ministry to turn down the application by Qualcomm, because of the demand to open up Japanese telecommunications markets that had been made by the US Trade Representative.

A solution was announced by Qualcomm in June 2000: the company would not apply for a 3G licence. One of the three Japanese mobile providers, a company in the DDI group, would use a CDMA 2000 technology supplied by Qualcomm. The details what happened behind the scene have not been disclosed. It is understood, however, that MIC made some adjustments with DDI for using the US technology in consideration of US–Japanese relations. Further, this may have been supported by the fact that the then president of DDI was a former deputy director of the Ministry who had retired a few years earlier.[20]

c. THE CASE OF SBB AND THE 800 MHZ BAND, 2003–

The 3G mobile service actually started in 2001. Its growth was slow in the beginning, but it is gathering speed currently. By December 2004, 10–25 per cent of the 85 million mobile subscribers in Japan were using IMT-2000 (the percentage changes depending on the definition of the boundary to be drawn between 2G and 3G). The price of 3G is falling because of scale economies and competition; it is expected that a dominant proportion of the mobile subscribers in Japan will choose 3G within a few years.

MIC assigned some of the 2 GHz bands to each of the three providers for 3G mobile. To accommodate increasing number of subscribers to 3G, MIC is planning to open up additional bands at 1.5 GHz and 1.7 GHz, with a forecast that the supply of spectrum for 3G will soon be in short supply.

MIC is also planning a transition of spectrum use from 2G to 3G. The project for reorganizing the 800 MHz band is one such plan. The 800 MHz band has been used intensively for 2G, since its technical properties are suitable for mobile telephony under presently available technology. For historical reasons, however, the current allocation of the 800 MHz band is fragmented, causing inconvenience and inefficiencies. In 2000, MIC started a project for reshuffling it. In August 2004, MIC invited public comments on a draft plan for assigning 3G blocks in the 800 MHz band. Just a few days before the deadline, Softbank Broad Band Corporation (SBB), a rapidly growing provider in telecommunications known for tough business conduct in wireline broadband communication, disclosed that the company would seek entry into the 800 MHz band so as to start a 3G mobile business. MIC informally rejected the request on the grounds that the 800 MHz project was for reorganization only and that the band was to be assigned to the three incumbent providers. SBB protested against this decision, and conducted a major campaign using advertisements in daily newspapers, requesting SBB customers to send their comments to MIC by email. It is reported that some tens of thousands of comments were sent to MIC by the deadline.

This has started the first case in Japan in which spectrum assignment with excess demand is being considered openly by MIC. That is to say, for the three 3G licences in the 800 MHz band there are now four applicants: the three incumbents and SBB.

MIC started a couple of informal study groups to seek a solution. Since October 2004, the study groups have met many times, but as of January 2005 no conclusion or solution had been reached. By the way, news on such matters is reported frequently in periodicals specializing in telecommunications (such as *Nikkei Communications*) and on the Web, but it is rarely reported in daily newspapers or in television news.

Possible outcomes of this case are as follows:

1 MIC may persuade one of the three incumbents to withdraw from the 3G market, but to do so would be difficult.
2 MIC may arrange for SBB to cooperate for 3G with an incumbent (such as starting a joint venture); this seems possible but not so easy.

3 MIC may reorganize the 800 MHz band so as to accommodate four 3G licences. This is not impossible, but one of the technical study groups would have to rework 800 MHz, which would be quite unusual. Further, MIC engineers would strongly oppose such a reorganization.
4 The three licences may be put up for auction. It would take a few years for MIC to prepare an auction, though this may be a solution in the long run. We still do not know what will come of this case.

3.3.2 Digital television

Transition of terrestrial television broadcasting from the current analogue system to a new system, digital television started in Japan at the end of 2003, several years after DTV was started in the United Kingdom and United States. Broadcasters in Japan, enjoying monopoly profits in a market shielded from a new entry, were not so eager to change over to the new system. MIC, however, applying its power to the broadcasters, successfully persuaded them to agree to the transition to DTV. According to a plan disclosed by MIC, the changeover will be completed in 2011; the assignment of spectrum to analogue TV will then be ended. It is not certain, however, whether this deadline will actually be enforced in event that there remains a large number of analogue viewers in 2011; no law or rule has been written on this point yet.

In Japan, the cost of moving over to DTV is to be borne by broadcasters and viewers (consumers), although MIC has arranged subsidies to broadcasters located in non-urban areas with unfavourable business conditions. The total investment to be made by broadcasters for DTV is approximately ¥1.2 trillion (i.e. US$100 per person), which is not so difficult to finance, since there is an abundant excess supply of funds in Japan and the long-term interest rate is as low as 2 per cent per year.

There is one complication in reassigning spectrum for the DTV transition in Japan. First of all, during the transition period, television programmes need to be broadcast on both analogue and digital channels (simultaneous broadcasting), since it takes years for all viewers to replace their TV receivers from analogue to digital. Also, television as well as telephony is considered to be a universal service in Japan. Thus, virtually anybody living anywhere in Japan should be able to view at least two or three terrestrial television channels. In an attempt to satisfy this requirement, Japanese broadcasters have constructed tens of thousands of television antennas (mostly for programme relaying) throughout the country, with subsidization by MIC. Since Japan is so mountainous, virtually all the 60 television channels, each of 6 MHz width (i.e. spectrum bands allocated to TV in the VHF or UHF radio frequencies) are used to deliver analogue TV programmes to every house without interference. Few extra channels are left available for the DTV transition. Simultaneous broadcasting seemed difficult in Japan.

Fortunately, however, recent technology has made it possible to reshuffle a small number of television channels in a particular region by temporarily moving a few analogue broadcasts frequency-wise, and accommodating DTV broadcasts on the vacated channels. By repeating this, all analogue channels could be converted to digital in each region. This is called the 'analogue-to-analogue transition' in Japan, and it started in 2003. It is forecast that the DTV deployment will be completed throughout the country by the end of 2006. The cost for this is approximately ¥200 billion, which is to be paid from the spectrum user fees borne mainly by mobile subscribers. (The rapid growth of mobile telephony from the second half of the 1990s onwards brought huge windfall-type of revenue to MIC (see p. 136). MIC is carrying out this policy despite opposition and critiques by experts and scholars concerned at what they see as the unjustifiable use of revenue. This is an example of MIC using its regulatory power in order to promoting a project it is determined to conduct, such as the DTV transition.

3.3.3 Wireless LAN

a. OUTLINE

Wireless LAN provides wireless access to the Internet. This is an example of which the technology was developed and the device manufactured and marketed before any spectrum was allocated for it. Initially, wireless LAN was started by using an unlicensed band. Within a few years, the demand for wireless LAN increased rapidly because of its usefulness to Internet subscribers, and the spectrum for it became congested. Internet users, including academics, began the 'freedom of spectrum' campaign internationally, which called for removal of all regulations such as licensing from the use of spectrum in order to establish 'spectrum commons' for wireless LAN. The ITU did not accept this request, but it recognized the shortage of spectrum for wireless LAN and allocated a significant amount of spectrum in the 5 GHz band for it. On the other hand, the technology for wireless LAN was expanded from indoor use to outdoor use, and wireless LAN became a new business in the 2000s.

The demand for wireless LAN has been strong in Japan, as in other countries. Unfortunately, however, the 5 GHz band allocated for wireless LAN by the ITU has been assigned to other uses, mainly for radar, and cannot be used for wireless LAN in Japan. MIC made an emergency arrangement to free a small portion of the 5 GHz band to be used for wireless LAN. Further, an additional 100 MHz in the 4–5 GHz band was released in 2004 by terminating the use of the spectrum by (wireline) telecommunications operators before the expiry of their licences. As a consequence, wireless LAN providers may use the band starting in 2005 but with additional spectrum fees for ten years, to bear half of the cost of compensation to be paid to the telecommunications operators for early termination of their licences (see p. 133).

This solved the urgent need for spectrum for wireless LAN. It is expected, however, the demand for wireless access to the Internet will increase significantly in the future as the Internet itself grows. For example, a service for wireless IP telephony is about to start; this may become a strong competitor to 3G mobile telephony. A significant shortage of spectrum for wireless LAN may emerge in the near future.

b. THE CASE OF MOBILE INTERNET, INC., 2002

Mobile Internet, Inc. (Mobile) is a venture business providing outdoor wireless LAN services. Mobile started its service in 2001 by constructing base stations for outdoor wireless LAN at a number of business and commercial centres in the Tokyo metropolitan area. A base station for outdoor wireless LAN covers an area within a diameter of 100 metres or so, which is called a hotspot. In 2002, Mobile planned to open hotspots at a few railway stations in Tokyo operated by Japan Railroad East, Inc. (JR), which is a private corporation operating under a special law; Mobile requested JR to let the company use a space within the railway station for wireless LAN. Mobile intended to lay down cables connecting a hotspot to one of its operation centres located outside the railway station. JR rejected Mobile's request, partly for the reason that JR itself was planning to start wireless LAN services.

Mobile appealed for relief to MIC on the ground that an article in the Telecommunications Business Law in Japan grants a telecommunications carrier (such as Mobile) the right of way on a property (real estate) owned by a private party in order to construct facilities needed to provide a telecommunications service (such as wireless LAN access) to the general public. MIC, after consulting with the Dispute Resolving Commission, rejected it by stating that, in effect, the right-of-way privilege was originally enacted for a long-distance telecommunications carrier to use a private property to construct trunk lines; the privilege is not for enabling a new business to provide an Internet access service on private property owned by a third party. Many experts expressed an opinion supporting Mobile, not MIC, in this case. Mobile, however, filed no further appeal for this case.

This made it clear that there are problems in Japan around the right of way for communications. Mobile, in appealing to MIC, used only the right-of-way argument as enacted in the Telecommunications Business Law, since this was the only article that Mobile could use in seeking to open wireless LAN stations on JR's space. The meaning of the article, however, is not clear; even the Dispute Resolving Commission recognized this and stated that further legislation would be needed to finally resolve the case of Mobile. Furthermore, it is unclear, under the present law, aside from the issue of the right of way, whether or not a provider has a right to use a spectrum band that is open to the general public (such as unlicensed bands, or bands that can be used with blanket registration) on a property owned by a third party.

3.4 Reallocation of spectrum, 2001–

3.4.1 Outline and major policy reviews

By 2000, it had become clear that the supply of spectrum was significantly short of the demand arising from growing services such as mobile telephony and wireless LAN, not to mention potential new services that could be made available by the development of digital wireless technology. To address this, MIC would have to engage in at least two tasks: (1) to create additional supply of spectrum by releasing some of the spectrum bands being used inefficiently (two steps are needed to do this: where to find such spectrum bands, and how and to where to relocate incumbent users); and (2) to allocate and assign the released spectrum for new use (MIC needs to decide whether to use C/C or a market mechanisms such as an auction to do this).

This subsection is devoted to summarizing the recent activities of MIC on this. In short, MIC has done the following:

1 to find out the current state of spectrum utilization and disclose it to the public
2 to evaluate the degree of efficiency of spectrum utilization for each band or each block
3 to establish a regulatory procedure through which inefficiently used spectrum is to be vacated; and
4 to reallocate or reassign vacated spectrum for new uses.

It is expected that by the end of 2005, a system for items 1 to 3 above will be prepared, including amending the Radio Law. Regarding item 4, I can make no forecast at the time of writing, except to say that a 100 MHz band in the 5 GHz area is to be reallocated to wireless LAN in 2005.

In January 2001, a group of academics specializing in spectrum policy published a proposal that MIC should disclose more information about the present state of spectrum utilization, and also that MIC should consider using a market mechanisms such as an auction in place of C/C for spectrum assignment, on the grounds that spectrum would be used with more efficiency and equity under a market mechanism than under C/C. The proposal explained further the overall state of current spectrum utilization. In the past, spectrum allocation and assignment were made on the assumption that there was an ample supply of spectrum to satisfy old and new needs. Spectrum was allocated and assigned whenever a new spectrum became usable thanks to technological progress, and was given to users without competition. No charge was imposed to users, except fees to cover the cost of administration and regulation. Further, licences are almost always renewed upon expiry. This policy has continued for about a century, beginning with the use of LF and MF (long and middle waves) for Morse telegraphy and voice radio, and continuing with the use of higher frequencies such as VHF and UHF for television and other uses. As a consequence, some spectrum is used very efficiently, in cases where the service was expanded rapidly (such as those for mobile telephones), whereas, as already mentioned, there are many other frequencies that are being used inefficiently, including some that area near idle. This may be compared to a scene in which skyscrapers are located next to rice fields. In the case of land space, market forces exclude the emergence of such an unbalanced use, but in the case of spectrum, no market forces are in action, thus leaving an unbalanced utilization to continue indefinitely.

In the course of planning and executing spectrum reallocation, MIC conducted some major reviews of its policy to make its intentions clear to itself and to the general public. Those reviews are summarized in the following.

In 1996, shortly after a series of major auctions for 2G licences (PCS auctions) had been conducted in the United States, MIC established a study group to examine the advantages and disadvantages of auction as a means through which to assign spectrum blocks. A report of the group was disclosed in 1997, stating that auction would indeed bring some advantages, including: (1) giving spectrum blocks to those who can best use them; (2) shortening the time needed for licensing; and (3) making the assignment process more transparent and thus decreasing uncertainty among potential users. But the report stated that, at the same time, a change of policy would have disadvantages, including: (4) assigning spectrum only to the rich; (5) letting some users monopolize spectrum; and (6) increasing the price of services supplied on spectrum. The report concluded that in view of the advantages and disadvantages as stated above, further investigation would be needed to decide whether or not to use an auction mechanism for spectrum assignment. The report also stated that at the time the report was written, auction was not worth considering in Japan, since there was no spectrum to be assigned anew. This statement, however, was false, since an assignment of 3G spectrums took place a couple of years after the report was written.

In 2000, MIC initiated a sequence of policies for reallocating spectrum systematically. In that and the following years, as is widely known, 3G auctions in Europe, particularly those in the United Kingdom and Germany, resulted in amazingly high bids for 3G licences. It is understood that MIC, on learning of this outcome in Europe, decided firmly that auction was out of question for spectrum assignment in Japan, since excessively high spectrum prices would bring a number of undesirable consequences such as delay in service deployment and financial difficulty for mobile operators. Thereafter, MIC has repeatedly published this view in reports written by study groups. These reports, however, did not discuss the implications of spectrum auctions conducted in the United States.

In 2003, MIC asked the Telecommunications Policy Council to write a report entitled 'Radio Policy Vision'. This is essentially an outline of the policy MIC had been conducting since 2000. The report summarized the increased use of spectrum during recent years and pointed out the current and future shortage of spectrum. The report further stated an outline of MIC's plan for reallocating spectrum, in addition to other topics on reforming spectrum user fees, promotion of R&D, and securing a safer environment for spectrum use.

3.4.2 Disclosure of information about spectrum allocation and assignment, 2000 and 2002

Traditionally in Japan, the Radio Department of MIC has disclosed little information on spectrum. For example, until 2001 the annual MIC white paper had no pages devoted to the subject.

Thus, MIC's first task in reallocating spectrum is to prepare a good information environment for incumbent and future spectrum users as well as for MIC itself. In fact, to disclose information on the present state of spectrum is a necessary condition for MIC to obtain users' agreement to its reallocation plan, regardless of whether they are asked to cease using spectrum altogether or to move elsewhere and continue using spectrum. Without such information, extremely strong resistance by outgoing or moving users would make a reallocation plan impossible to execute.

The task was conducted in two steps: (1) disclosure of information about spectrum allocation; and (2) disclosure of information about spectrum assignment (i.e. information about licences).

Disclosure of information about spectrum allocation was initiated in 2000 when an amendment of the Radio Law was agreed upon in Parliament. In particular, a framework called 'Spectrum Assignment Plan', which really means a spectrum allocation plan in our terminology, was implemented, together with MIC's obligation to disclose it (RL26). Until that time, no such formality existed, except in a simple form. The plan is a table of spectrum bands that are being assigned to users, or which are not assigned and can be assigned to new users. The plan is useful particularly to potential users of spectrum.

Disclosure of information about individual licences, including disclosure on the Internet, was conducted according to an amendment of the Radio Law in 2002 (RL25). This is really a big change in spectrum regulation in Japan, since prior to that, nothing was disclosed about individual licences. Careful screening of the items of a licence to be disclosed was done in such a way that the disclosure gives ample information about the state of spectrum frequencies being used, so as to benefit incumbent and potential users, but does not invite undesirable intervention or interference that would damage the convenience and the privacy of incumbent users unnecessarily. The information disclosed should be a good source from which to compile useful statistics; no statistics concerning spectrum use, however, have yet been published commercially or by MIC.

3.4.3 Review of spectrum utilization, 2002–

As stated previously in this chapter, MIC is determined to use C/C for spectrum allocation and assignment. Under C/C, it is necessary for MIC to know the degree of efficiency with which each spectrum band or each spectrum block is used in order to conduct reallocation of spectrum. If some market mechanism (such as competitive lease of a licence) were used, it would be unnecessary for MIC to assemble such information, since reallocation would be achieved as a consequence of market transactions. MIC, however, did not take this course and accepted the burden of information collection and processing, together with the trouble of justifying a reallocation plan to users.

Thus, the Radio Law was amended in 2002 to give MIC the authority to conduct a systematic survey and evaluation of spectrum uses with individual licences (RL26-2). The law requires that, usually, such investigation be made for each licence once every three years. Spectrum users must report, by filling a form prepared by MIC, on the use of spectrum such as the amount of traffic, the time of day when a station is active, and other details of station operations.

After collecting the reports, MIC is supposed to evaluate the efficiency in which a spectrum band or a spectrum block is used. An outline of the evaluation by MIC is to be disclosed, but not its details. Later in 2004, the Radio Law was again amended to give MIC the authority, based on the outcome of evaluation, to terminate a licence even before its expiry date. Thus, MIC has obtained a strong power to conduct spectrum reallocation (see the next subsection).

In 2002, the first survey of spectrum use was conducted. An outline of the survey, together with MIC's evaluation, was disclosed in 2003. Likewise, surveys and evaluations were conducted in 2003 and 2004.

3.4.4 Reallocation with compensations, 2003–

In January 2002, MIC started a study group on policies concerning the effective use of radio spectrum. This study group discussed, through to July 2004, a number of subjects, including spectrum reallocation and expansion of spectrum user fees. One subject was the creation of a system for spectrum reallocation by which spectrum users whose licence is terminated before its expiry date would be suitably compensated.

Let us explain this system in more detail. First of all, the Radio Law formally gives MIC the authority not to renew a spectrum licence, although spectrum licences are usually renewed. In earlier years, when spectrum shortage was not so serious, spectrum reallocation was executed with caution by giving plenty of time (such as ten years or more) to users to prepare for a transfer to another spectrum band. Table 5.1 lists some of the major spectrum reallocations in Japan. Many of them were conducted according to a decision made by an international agreement in the WRC. In summary, there had been cases of spectrum reallocation, but it had been very rare for MIC to use its power to terminate a licence even at its expiry.

When the shortage of spectrum became serious, as in recent years, MIC decided to use its power not to renew a licence upon its expiry, by first making clear to spectrum users that licences need not be renewed automatically. In particular, MIC made it clear that, since a licensee did not have the right to continue using a block beyond the licence term, no compensation would be given when a licence was terminated for reallocation at its expiry (i.e. if it was not renewed). However, if licence is terminated in mid-term, some compensation should be paid to the user.

The study group investigated, on this principle, how much compensation should be paid when a spectrum user is taken off a licence before it expires. The study group reported in 2003 that a compensation amount should be equal to the value of the user's tangible capital stock (such as a wireless tower or equipment) that had not yet been depreciated with allowances. Other assets, including intangible ones, such as workers' wireless skills, should not be counted for compensation. Thus, MIC adopted a very strict rule in calculating compensation for reallocation. This report was enacted in 2004 to amend the Radio Law (RL71-2).

The study group also investigated sources of money for paying such compensation. The group recommended that half of the compensation be paid out of spectrum user fees collected by MIC, and the other half be paid by the new user to whom the reallocated spectrum is assigned (RL103-2). Furthermore, it is also recommended that the selection of a new user may be made competitively according to the amount of contribution that a user offers to be used for compensation in order to obtain a licence. MIC calls this a 'quasi-market mechanism for selecting spectrum users'.

In 2004, MIC actually executed a reallocation of spectrum in the 4.9–5 GHz band, which had been used by telecommunications operators; it is understood that radio spectrum will be replaced by optical fibres. It is also understood that, so far, the cases where MIC has been successful in actually relocating spectrum users are those in which users are telecommunications common carriers, operating directly under regulation by MIC. It remains to be seen whether or not MIC will be able to relocate other users such as power-line operators from their current use of spectrum.

Table 5.1 Relocation of spectrum bands in Japan, 1965–2007

Relocated bands (MHz)	*Year*	*Objective of use before relocation (moved to:)*	*Objective of use after relocation*	*Reason for relocation*
2, 27, 150	1965–1966	Ship station on board Coastal fixed (150 MHz band)		Inland navigation
2.85–6	1965–1968	Aeronautical fixed		International agreement at WRC
10–17	1966–1969	Aircraft station on board		International agreement at WRC
6–8	1966–1970	Designated use (HF)		International agreement atWRC
4–2.3	1967–1970	Ship station on board Coastal fixed (HF)		International agreement at WRC
4-2.3	1974–1978	Ship station on board Coastal fixed (HF)		International agreement at WRC
4–2.75	1987–1991	Ship station on board Coastal fixed (HF)		International agreement at WRC
60	1997–2007	Fixed local government use for safety (400 MHz band)	Fixed local government use for safety	Increased demand
33.54–470	1982–1995	Fixed multiplex (2 GHz band)	Mobile government use for safety	Increased demand
	1995–	Fixed multiplex (3 GHz band and above)		
940–960	1981–1992	Fixed for broadcast operation (3.4 GHz band)	Mobile telephony	Increased demand
1,500	1985–1991	Mobile government use	Mobile telephony	Increased demand
2,000	1992–2002	Fixed public use (6.5 GHz, 7.5 GHz bands) Fixed telecommunications use (4 GHz, 5 GHz, 6 GHz bands)	IMT-2000	1992 WRC for IMT-2000
21,000–22,000	1992–2002	Mobile telecommunications use (22–22.4 GHz, 22.6–23 GHz)	Satellite broadcast	1992 WRC for reorganization of 20 GHz band
23,000–23,200	1992–2007	Fixed for cable TV operation (Fixed: 23.2–23.6 GHz) (Mobile: 21.2–21.4 GHz)	Fixed for telecommunications operation	1992 WRC for reorganization of 20 GHz band

Source: Compiled and translated by the author from MIC publications.

3.5 Spectrum user fees

3.5.1 Outline

There are two types of fees that spectrum users pay in Japan. The first is the processing fee, which covers the cost of processing licence applications, equipment certificates, etc. The revenue from this fee goes to the general budget of the Japanese government. The second is the spectrum user fee, first introduced in 1992. The user fee was considered to cover the cost arising from spectrum regulations. Furthermore, the revenue from this fee goes directly to the revenue of a budget which MIC can spend how it thinks best (RL103-3); the idea behind this was that since the user fee is spent directly on the cost arising from MIC's regulatory activities, the revenue need not be included in the general budget of the Japanese government. The budget that is administered directly by MIC is called a 'specified budget'.

At the time that the user fee was introduced in Japan, spectrum fees in some form or other had been popular in many countries, including the United Kingdom, Germany, France, Canada, Australia and New Zealand. Further, a bill to introduce regulatory fees (including but not limited to spectrum regulation) was proposed to Congress in the United States. These helped the Ministry explain the need for spectrum fees to Parliament.

Once the user-fee system started, MIC utilized this framework to increase its revenue repeatedly. As a consequence of this, and also helped by the increase in mobile phone subscriptions, a large amount of revenue from spectrum user fees is flowing into MIC's budget today. On average, a mobile subscriber in Japan pays MIC, through mobile operators, ¥540 a year, approximately one hundred times greater than the average amount that a US mobile subscriber pays through operators to the FCC as a regulatory fee, although in the United States, mobile subscribers using PCS and the service thereafter bear the cost of auction payment that an operator has remitted to the FCC to obtain a licence.

3.5.2 Initial introduction of the spectrum user fee, 1992

An amendment of the Radio Law in 1992 (RL103-2) introduced a spectrum user fee for the first time in Japan. It is stated in the law to be 'the money that licensees shall bear in order to pay the expenses of administrative work, including the following, with the objective of benefiting all radio stations as a whole concerning the assurance of the sound use of radio waves (RL103-2(3))'. Specifically, the Radio Law as of 1992 listed two objectives: (1) monitoring the use of radio spectrum; and (2) maintaining a database of spectrum licences. The fee for various types of wireless stations is also specified in the law; a mobile subscriber should pay ¥600 and a broadcaster ¥29,700 annually to MIC.

The basis for calculating these annual fees was explained by the Ministry as follows. First, the benefit that each user receives from monitoring is spread evenly over all stations; therefore, an equal amount should be borne by all users. Second, the cost of the database is approximately proportional to the length of a record storing the licence information of a user; therefore, the total cost of the database should be borne by users in proportion to the length of the record. The Radio Law did not state, and the Ministry did not explain, however, how the amount paid by a user would be divided between the two objectives.

It was stated by the Ministry, but not written into the Radio Law, that the user-fee scheme should be re-examined every three years for possible revision to reflect changes in the cost of administration.

3.5.3 Expansion of user-fee revenues and expenditures, 1996–2003

After the introduction of spectrum user fees, the growth of mobile telephony started; the number of mobile phones grew from around one million in the early 1990s to 80 million in 2004. The revenue from spectrum user fees grew enormously as a consequence of this. Today, more than 80 per cent of the revenue from spectrum user fees is paid by mobile telephone subscribers. This is an outcome of the fact that a fixed amount of user fee for each licensee is written into the law, and no attempt has been made either by MIC or by Parliament to adjust it according to the increase in number of mobile subscribers. Many experts outside MIC have been disturbed on learning these statistics. However, it has not become a major talking point in Japanese society, partly because the amount of the fee borne by a mobile subscriber is low relative to the total spending on mobile phones (1 per cent or so), and also because the fee is not paid directly by mobile subscribers, but paid through mobile operators instead. Further, fees imposed on broadcasters have been low relative to the level of their operation; there is no incentive for broadcasters to make spectrum user fees a news item on the media.[21]

In accordance with the increase in the revenue from spectrum user fees, MIC added, by revising the Radio Law one clause after another, new objectives for which the revenue could be spent.

1. In 1996, a new objective of ‘testing and analysing wireless devices’ was added.
2. In 1997, the system of blanket licensing was introduced so that information concerning licences of mobile subscribers would no longer be included in the database. In spite of this, the Radio Law was amended so that a mobile subscriber would continue to pay a fixed amount of ¥540 a year (10 per cent lower than the original ¥600). This contributed greatly to the increase in the revenue collected as spectrum user fees.

3 In 2001, an additional objective of using the revenue was added, which was the cost of reassigning spectrum for the introduction of DTV.
4 In 2003, the fee to be collected from a broadcaster was increased significantly, but for the coming ten years only, in consideration of complaints by mobile operators that MIC was using the revenue collected mostly from them in order to subsidize the introduction of DTV.
5 In 2004, further legislation amended the Radio Law to the effect that the revenue from spectrum user fees may be used to pay half the cost of compensating incumbent users whose licence was terminated for spectrum reallocation (see p. 133).

Through the revisions listed, the nature of the spectrum user fee has been changed from paying the administrative costs of maintaining sound use of spectrum to conducting a number of spectrum policies as prescribed by MIC. Thus, the spectrum user fee today, in effect, should be regarded as a tax, rather than as a fee for recovering the cost of public services (spectrum administration in this case) supplied by MIC to users, despite the fact that the Radio Law still keeps the same wording defining 'the spectrum user fee' as was originally introduced in 1992. MIC has been attempting to justify this change in the nature of spectrum user fees; but its explanation remains ambiguous.

3.5.4 Introduction of a 'spectrum user fee based on congestion', 2004–

In 2004, a plan was in progress to introduce yet another spectrum user fee that depends on the degree of congestion in spectrum utilization. The study group, having released a report on spectrum reallocation at the end of 2003, was asked to discuss the issue of introducing this new fee. The final report of the group was made available in October 2004, after public comments had been collected. It is understood that MIC will soon be writing a bill to amend the Radio Law in the light of this report, at the time of writing.

According to the final report, a new spectrum user fee will be imposed based on the degree of congestion of spectrum use. The amount of the fee for each licensee will be determined, therefore, according to the band of frequencies and the size of spectrum blocks being used, as well the location and the size of the area of use (i.e. the emission power). MIC has attempted to emphasize in the report that the arrangement resembles the economic price of spectrum. In fact, however, no market mechanism, such as demand–supply adjustments, is included in the plan; the actual fee rate would be set by MIC at a level far below the 'equilibrium price' level in a congested area. Consequently, the new fee will not bring in the benefits of market mechanism, except that in cases where spectrum utilization is extremely inefficient in a congested block, the new system might work to drive out the inefficient user. Thus, this new fee resembles an excise tax rather than an economic price.

It has been reported that MIC is considering imposing the new spectrum fee on users who share a spectrum block, including wireless LAN users, new RFID users and other users of unlicensed bands. Strong opposition to this plan was raised in the fall of 2004; it has been reported that MIC may be abandoning this plan by limiting the introduction of the new fee only to licensees, not registered users or users of unlicensed bands.

4 Characteristics of spectrum regulation in Japan

Spectrum regulation in Japan is considered to differ significantly from that in other countries. Most of the characteristics of spectrum regulation in Japan, in fact, are shared by regulations concerning transport, power supply and many others. The behaviour of MIC as a regulatory agency resembles that of other ministries. In other words, these are an outcome of the structure and the function of Japanese government and the Japanese society. For this reason, it seems difficult to characterize spectrum regulation in Japan to its full extent.

In this section, therefore, an attempt will be made to 'answer' a question on spectrum regulation in Japan: 'Why has MIC insisted on continuing to use C/C for spectrum allocation and assignment?' In many countries other than Japan, advanced or not, there is a trend to use market mechanisms such as auctions for spectrum assignment. All MIC officers know this, of course. Economic theory points out the shortcomings of C/C as a means of regulating spectrum in the age of its shortage. Many MIC officers understand the theory. There must be reasons why MIC continues using C/C. What are they?

There are two conventional answers to the question. The first is that if auction or some other market mechanism were introduced, MIC would lose its 'administrative power' generated from supplying spectrum resources exclusively at its discretion. The second is that if auction or some other market mechanism were introduced, the auction revenue would go to the Ministry of Finance (MOF) according to the present budget scheme, not to MIC; that is, MIC would lose its revenue. Let me further explain these answers.

First of all, regarding the power of MIC, it is easy to understand that MIC's supply of spectrum under C/C gives some power to control spectrum users, since they fear losing the right to use spectrum if they resist MIC in any way. The true question, therefore, may be the following: 'Why does MIC need such power for regulation?' Further, 'why does MIC want the power that derives from retaining C/C for spectrum regulation? Is it not the case that a regulatory power should be given by law, not by the supply of spectrum?'

With regard to the revenue from auction going to MOF, a question may be asked: 'Does it make any difference whether the revenue goes to MOF or is retained by MIC?' The answer is YES. Still another question may follow: 'Is it not the case that all revenues to the government should be administered under a central authority, which distributes the money to all ministries according to their needs under the government budget system?'

The reader may say, 'All bureaucratic systems seek power and money. How come only in Japan has C/C been retained by MIC because of the power and money it confers?' The answer to this question is that the difference between Japan and other counties with regard to the use of C/C for spectrum assignment lies in the difference in the need to seek power and money, and in the capability to obtain power and money, between MIC and bureaucrats in other countries. Before discussing these two points, we need to examine the functioning of the government ministries in Japan.

First, each ministry of the executive branch of the Japanese government is more or less an independent entity. The authority of the prime minister and the cabinet is weak relative to other countries. For instance, the prime minister can nominate and fire a cabinet member, but it is not so easy for him or her to do the same thing with a ministry officer, even indirectly. A ministry has the responsibility to support its members, particularly career officers. To this end, a ministry has a strong incentive to increase its budget by using its power. Under the present system, a ministry is considered to be informally responsible for supporting even those members who retire earlier than their colleagues; thus, the ministry uses part of its budget to create public organizations for various works closely related to the ministry's activity, so that retired ministry officers may be re-employed there. The need to secure such re-employment of retired officers is strong in Japan because of its lifelong employment system.

Thus, the need for power and money is strong in all ministries. For MIC, spectrum regulation is a good source of power and money as long as C/C is retained.

To avoid possible misunderstanding, let me emphasize that what I have just said does not imply that ministry officers in Japan always seek self-interest rather than serving to the people for the public interest. Nor does it mean that government officers are more selfish than non-government workers. My comments concern the way in which government officers are organized in a ministry, not how they work, or intend to work. For the case of MIC, as we saw in a preceding section, spectrum regulation carried out by MIC has been quite good in quality within the framework of C/C.

Notes

My gratitude goes to the editors of this book, Professors Ruth Taplin and Masako Wakui, for their advice, encouragement and patience given to the author while this chapter was being written.

1 Paul Milgrom, *Putting Auction Theory to Work*, (Cambridge: Cambridge University Press, 2004).
2 FCC (US), *Spectrum Policy Task Force Report*, ET Docket No. 02-135, released in November 2002 (http://www.fcc.gov/sptf/reports.html, http://hraunfoss.fcc.gov/edocs_public/attachmatch/DOC-228542A1.pdf).
3 An example of spectrum allocation in the form of a map may be seen with 'US Frequency Allocation Chart', National Telecommunications and Information Administration (NTIA), US Department of Commerce (http://www.ntia.doc.gov/osmhome/allochrt.html).
4 In this chapter, the two terms, *allocation* and *assignment* will be used with the meaning stated in the text, since this is the standard usage worldwide. In documents issued by the Japanese government, however, the word 'assignment' is frequently used to mean allocation, whereas the word 'licensing' is used to mean assignment. Readers should be aware of this difference in terminology when examining a document coming from the Japanese government.
5 An exception to this is the advocacy for 'spectrum commons', which is explained at the end of this subsection.
6 However, spectrum reallocation by means of insurance and compensation has been proposed by the present author: Hajime Oniki, 'Reallocation of Radiowave Spectrum with a Price Mechanism: Proposal of a System of Insurance and Compensation', presented at the 32nd Research Conference on Communication, Information and Internet Policy (Telecommunications Policy Research Conference 2004) held at George Mason University, Fairfax, VA, 1–3 October 2004 (http://www.osaka-gu.ac.jp/php/oniki/noframe/eng/publication/200408a.html, http://web.si.umich.edu/tprc/papers/2004/367/Maintext.pdf).
7 See, among others, Gerald R. Faulhaber and David Farber, 'Spectrum Management: Property Rights, Markets, and the Common', 17 April 2002 (http://bpp.wharton.upenn.edu/Acrobat/Faulhaber_AEW_paper_6_19_02.pdf); Thomas W. Hazlett, 'Assigning Property Rights to Radio Spectrum Users: Why Did FCC License Auctions Take 67 Years?', *Journal of Law and Economics* 41 (2): 529–576 (1998); Evan Kwerel and John Williams, 'A Proposal for a Rapid Transition to Market Allocation of Spectrum', OPP Working Paper Series No. 38, Federal Communications Commission, Office of Plans and Policy, 2002, pp. 1–54 (http://hraunfoss.fcc.gov/edocs_public/attachmatch/DOC-228552A1.pdf).
8 See, for example, Paul Baran, 'Is the UHF Frequency Shortage a Self Made Problem?', speech given at the Marconi Centennial Symposium, Bologna, Italy, 23 June 1995 (http://www.interesting-people.org/archives/interesting-people/199507/msg00023.html); Nobuo Ikeda, 'The Spectrum as Commons: Digital Wireless Technologies and the Radio Policy', RIETI Discussion Paper Series, October 2002 (http://www.rieti.go.jp/jp/publications/summary/02030001.html), and papers cited therein.
9 See, for example, Benkler Yochai, 'Some Economics of Wireless Communications', *Harvard Journal of Law and Technology* 16 (1): 1–59 (2002) (http://jolt.law.harvard.edu/articles/pdf/v16/16HarvJLTech025.pdf).
10 Most of the reference materials used for this chapter are published only in Japanese. For a list of them, see the bibliography section of Hajime Oniki, *Economics of Spectrum Resources: U.S. Spectrum Auctions* (Sagamihara, Kanagawa, Japan: Gendai-Tosho, Inc., 2002) (in Japanese) (http://www.osaka-gu.ac.jp/php/oniki/noframe/jpn/publication/200202a.html).

11 Termination of a licence may not be extraordinary in the near future, though (see p. 132).
12 It is possible, however, for a non-Japanese corporation to obtain a licence through a Japanese subsidiary.
13 Law 108 of 28 April 1952.
14 It may appear that the communications part of MIC, as shown in Figure 5.1, is not organized logically, or that some of its divisions are not named properly. It may not be apparent why the broadcast divisions are within the Information and Communications Policy Bureau, and why the Radio Department is placed within the Telecommunications Bureau, which is supposed to handle wireline regulations, as the name 'Telecommunications Business Department' might suggest. This, however, is a consequence of 'reforms' imposed on the Ministry in the past; the actual division of work and responsibility within the communications part of MIC is basically the same as in other countries.
15 Note that such a procedure is seen more or less commonly with all ministries of the Japanese government.
16 After the occupation was ended in 1953, the Japanese government, in a trend called a 'reaction course', gradually replaced US-style systems with traditional Japanese-style ones. The trend covered a wide range of government activities and institutions, including spectrum regulation. This tendency, however, was reversed sometime in the late 1980s or early 1990s, as leaders in Japan started recognizing the importance of court-style institutions to conduct public administration under due process of law and in an open environment. For example, the Dispute-Resolving Commission was established by MIC in the 1990s for handling matters in wireline telecommunications.
17 In recent years, however, the Democratic Party, the major political party opposing the ruling LDP, has proposed a bill to establish a 'Communications Commission', a court-style regulator, to handle telecommunications regulation including spectrum. This party has also proposed a bill that would cause spectrum assignment to be conducted by auction. Since the Democratic Party is in a minority in Parliament at present, these attempts have not succeeded yet.
18 Economic theory states that this is one of the major reasons that the economy of the (the former) Soviet Union, which was run under C/C, encountered difficulty after the Second World War.
19 In Japan, it is understood that the law as a whole should maintain consistency. This is guaranteed by the fact that most bills are proposed to Parliament formally by the cabinet, which is responsible for maintaining legal consistency and integrity. This means, on the other hand, that the judicial sector is relieved of the work to solve inconsistencies in law. This also means that the responsibility (and the power) of the executive sector is strong in Japan.
20 DDI (KDDI today) is a private company started in the late 1980s as a new common carrier engaging in long-distance wireline telecommunications after the liberalization of the market in 1985. DDI as such is the only long-distance carrier that constructed its network mainly on microwave channels. It is likely that good relations had been built between MIC and DDI during the period when DDI was constructing its network.
21 The amount of spectrum user fee paid by a mobile subscriber in Japan is approximately 100 times greater than that in the United States, whereas that paid by a Japanese broadcaster is about one-fiftieth of that paid by a US broadcaster. See Section 9 of the Communications Act 1934 (US), which sets the regulatory fee for cellular or PCS radio at $60 per 1,000 subscribers and that for a VHF commercial TV at between $8,000 and $80,000 per market per year.

Appendix A: Major amendments to the Radio Law (listed in the order of article numbers)

Article no.	*Subarticle No.*	*Article header*	*Amendments*	*Year amended/ Law no.*
20	2, 3	Succession of a License	Succession of a licence allowed in the event that licensee juridical person is divided or sold	2000/109
25	1–3	Publication of Information	Information concerning radio stations disclosed to the public on the Internet	2002/38
25	1–3	Concerning Radio Stations	Information concerning radio stations created by registration to be disclosed	2004/47 Article 2
26	1, 2	Frequency Assignment Plan	Creation of a system of the frequency assignment plan	2000/109
26-2		Survey of Actual Radio Spectrum Usage	Creation of a system for surveying and evaluating actual radio spectrum usage	2002/38
27-2 to 27-11	*	Specified Radio Station, Blanket License	Creation of blanket licensing system	1997/47
27-12 to 27-17	*	Specified Base Station	Creation of blanket licensing for cell-type stations	2000/109
27-18 to 27-34	*	Spectrum Use by Registration	Creation of system for opening a station by registration (in place of licensing)	2004/47 Article 2
71-2	1	Specific Frequency-Change Support Service	Creation of a system for frequency relocation for specific purposes	2001/48
71-2	2	Specific Frequency-Termination Support Service	Creation of a system for frequency termination for specific purposes	2004/47 Article 1 and 2
76-2-2		Restriction of Registration	Give MIC authority to restrict opening a station by registration in case of congestion	2004/47 Article 2
76-3		Termination of License	Give MIC authority to alter or terminate a licence on evaluation of actual radio spectrum usage	2004/47 Article 1 and 2

103-2	*	Spectrum User Fee	Creation of spectrum user fee for paying the cost of handling licences and enforcing proper use of spectrum	1992/74
103-2	*	Spectrum User Fee	Testing and analysing wireless devices added as an objective of expense	1996/70
103-2	*	Spectrum User Fee	Fees under blanket licensing (a mobile phone user pays a fixed amount of ¥540 a year)	1997/47
103-2	*	Spectrum User Fee	Cost of specific spectrum change (reassignment for DTV transition, etc.) added as an objective of expense	2001/48
103-2	*	Spectrum User Fee	Cost of specific spectrum change adjusted with broadcasters	2003/68
103-2	*	Spectrum User Fee	Cost of specific spectrum termination (for reallocation) added as an objective of expense	2004/47 Article 1 and 2
103-3	1, 2		Establishes that user-fee revenues be used for spectrum regulation only (effectively excludes said revenue from being administered under the general budget of the Japanese Government)	1992/74

Source: Compiled by the author using materials published on web pages by MIC (http://www.soumu.go.jp/joho_tsusin/eng/laws_dt02.html, http://www.shugiin.go.jp/index.nsf/html/index_housei.htm).

Note
*indicates that sub-article numbers are not entered explicitly.

Appendix B: Chronology – Japanese spectrum policies, 1992–2004

Year	*Month*	*Day*	*Event*
1950	5	2	Enactment of Radio Law (1950Law131).
1992	6	5	Radio Law amended (103-2: Spectrum User Fee): Creation of spectrum user fee for paying the cost of handling licences and enforcing proper use of spectrum (1992Law74) (p. 135).
			Radio Law amended (103-3): Establishes that user-fee revenue be used for spectrum regulation only (effectively excludes said revenue from being administered under the general budget of the Japanese government) (1992Law74) (p. 135).
1996	5	10	The 1996–97 Study Group for Effective Radio Use started (p. 130).
	6	12	Radio Law amended (103-2: Spectrum User Fee): Testing and analysing wireless devices added as an objective of expense (1996Law70) (p. 136).
1997	2	4	MPHPT publishes a report from the 1996–97 Study Group for Effective Radio Use (p. 130).
	2	17	Nikkei reports that at least two CDMA standards, W-CDMA (Japan and Europe) and CDMA 2000 (US), would be used for IMT-2000 internationally; automatic worldwide roaming would be impossible. NC (1997.2.17, p. 168) (p. 124).
	5	9	Radio Law amended (27-2 to 27-11: Specified Radio Station, Blanket Licence): Creation of blanket licensing system (1997Law47) (p. 121).
			Radio Law amended (103-2: Spectrum User Fee): Fees under blanket licensing (a mobile phone user pays a fixed amount of ¥540 a year) (1997Law47) (p. 136).
1999	3	1	Nikkei reports that mobile web services such as i-mode of NTT DoCoMo were growing rapidly. NC (1999.3.1, pp. 98–117).
2000	3	27	MPTPT announces basic policies for introducing the IMT-2000 services, including that three spectrum blocks would be blanket-licensed nationwide (p. 124).
	5	1	A Japanese subsidiary of Qualcomm, Inc. (US) announced its intention to apply for a 3G licence. NC (2000.5.1, pp. 44–45) (p. 124).
	6	2	Radio Law amended (20: Succession of a License): Succession of a licence allowed in the event that a licensee juridical person is divided or sold (2000Law109) (p. 122).
			Radio Law amended (26: Frequency Assignment Plan): Creation of a system of frequency assignment plan (2000Law109) (p. 131).
			Radio Law amended (27-12 to 27-17: Specified Base Station): Creation of blanket licensing for cell-type stations (2000Law109) (p. 122).

	6	5	Qualcomm announces that it will not apply for a 3G licence; three providers, NTT DoCoMo, J-Phone and DDI, are expected to serve 3G. DDI would use a CDMA technology supplied by Qualcomm. NC (2000.6.5, pp. 51–53) (p. 124).
	10	20	MPHPT announced a Spectrum Assignment Plan (p. 131).
	11	6	Radio Technology Council reports that outdoor wireless LAN in the 5.3 GHz band cannot be realized because of interference with radars. NC (2000.11.6, pp. 86–87) (p. 127).
2001	1	26	A group of university scholars propose that auctions should be introduced for allocating spectrum. NK (2001.1.26) (p. 129).
	2	27	The Japanese government publishes ‘e-Japan Initiatives’, including a proposal to adopt spectrum auction. NK (2001.2.27) (p. 129).
	4	18	Radio Regulatory Council reports a draft of the ministerial ordinance in preparation for the introduction of terrestrial digital television broadcasting (DTV) (p. 126).
	5	28	MPHPT inquires of the Telecommunications Council about technical conditions for broadband mobile access system using 5 GHz band frequency (p. 127).
	5	30	NTT DoCoMo launches introductory service of IMT-2000 (FOMA) (p. 124).
	6	15	Radio Law amended (71-2: Specific Frequency-Change and Frequency – Termination Support Service): Creation of a system for frequency relocation for specific purposes (2001Law48) (p. 126). Radio Law amended (103-2: Spectrum User Fee): Cost of specific spectrum change (reassignment for DTV transition, etc.) added as an objective of expense (2001Law48) (p. 137).
	7	30	Study Group on Business Model over Next-Generation Mobile Communications System compiles a report.
	9	25	MPHPT receives a report from the Telecommunications Council on the ‘Technical conditions necessary for advancement of a wireless system using 2.4 GHz band’ (p. 127).
	10	1	NTT DoCoMo starts full service of FOMA, its IMT-2000 service. NC (2000.10.1, pp. 75–77) (p. 124).
	10	10	Sixth WP8F meeting of the International Telecommunication Union (ITU) held in Tokyo.
	11	26	MPHPT receives reports from the Telecommunications Council on ‘technical regulations for CDMA2000 1xEV-DO System’ and ‘technical regulations for the next-generation mobile communications system (CDMA/FDD) radio facilities’ (p. 123).
	12	12	Radio Regulatory Council reports on rule-making in relation to Law Concerning Broadcast via Telecommunications Carriers’ Facilities.
	12	25	Announcement of the report by the Research Group on Disclosure, etc. of Radio Spectrum Use (p. 131).
2002	1	28	Study Group on Policies concerning the Effective Radio Spectrum Use launched (p. 132).
	2	22	MPHPT submits a bill to amend the Radio Law, which contains survey and publication of actual radio spectrum usage and expansion of disclosure of radio station licence information (p. 131).

Year	Month	Day	Event
	3	4	NTT Communications, the long-distance carrier of NTT, launches its Hot Spot Service (wireless LAN access) at 500 locations in Japan. NC (2002.3.4, pp. 59–61) (p. 127).
	4	1	KDDI starts its IMT-2000 service, CDMA2001x. NC (2002.4.1, pp. 73–75) (p. 124).
	4	23	MPHPT launches the Power Line Communication Study Group.
	5	6	Nikkei reports that Mobile Internet, Inc. (Mobile, a wireless access provider), confronts Japan Railroad East Corporation (JR-East), following Mobile's request to use JR-East's space at major railway stations in Tokyo. NC (2002.5.6, pp. 43–5) (p. 128).
	5	7	MPHPT receives a report from the Telecommunications Council concerning 'technical conditions for 5 GHz band wireless access system' (p. 127).
	5	10	Radio Law amended (25: Publication of Information Concerning Radio Stations): Information concerning radio stations disclosed to the public on the Internet (2002Law38) (p. 131).
			Radio Law amended (26-2: Survey of Actual Radio Spectrum Usage): Creation of a system for surveying and evaluating actual radio spectrum usage (2002Law38) (p. 132).
	5	15	MPHPT submits inquiry to Radio Regulatory Council concerning partial amendments of Frequency Assignment Plan for the introduction of 5-GHz Wireless Access System (p. 127).
	7	1	MPHPT announces an interim report from the 'Study Group on Policies Concerning the Effective Radio Spectrum Use' (p. 132).
	7	29	Telecommunications Council starts investigations on matters including technical conditions for the effective use of mobile commercial-use frequencies in the 800 MHz range (p. 125).
	8	7	MPHPT announces an inquiry concerning development of Radio Policy Vision to Telecommunications Council on mid- to long-term outlook of radio spectrum use and the role to be played by the administration (p. 130).
	8	8	MPHPT rejects Mobile's request to use JR-East's spaces for providing wireless access services (p. 128).
	8	9	Announcement of report by Power Line Communication (PLC) Study Group, stating that immediate use of PLC would be impossible because of interference with existing systems such as amateur radio.
	9	27	Establishment of licence policies by MPHPT for terrestrial digital television broadcast stations. (p. 126)
	9	30	MPHPT inquires of the Telecommunications Council on technical conditions for ultra-wideband (UWB) radio systems.
	10	16	MPHPT receives a report from the Radio Regulatory Council on a draft of a ministerial ordinance concerning a survey of radio spectrum use (p. 132).

	11	22	MPHPT proposes to broadcasters that their spectrum user fees be increased approximately by seven times for spending on preparation of terrestrial DTV. NK (2001.11.22) (p. 137).
	12	20	J-Phone starts its IMT-2000 service. NC (2002.12.16, pp. 68–69).
	12	25	First report of the Study Group on Policies concerning the Effective Radio Spectrum Use announced (p. 132).
2003	1	27	MPHPT receives a partial report from the Telecommunications Council on the ‘Technical conditions necessary for advancement of wireless system using 2.4 GHz band’ (p. 127).
	1	30	Start of changing analogue TV frequencies for preparing terrestrial digital broadcasting (p. 127).
	2	4	Study group reported on the realization of a compensation scheme for the reallocation of the radio spectrum (p. 133).
	4	16	MPHPT receives a report from the Radio Regulatory Council concerning the decision on development plans for specified base stations and an amendment of the Frequency Assignment Plan – Looking ahead to the elimination of frequency shortage for cell phones (p. 132).
	4	16	MPHPT receives report from the Radio Regulatory Council concerning blanket licences for eight specified radio stations including NTT DoCoMo Inc. – Developing IMT-2000/PDC dual-mode terminals nationwide (p. 124).
	4	18	MPHPT receives a report from the Radio Regulatory Council that pre-permits shall be granted to terrestrial digital television broadcasting stations for which applications for licences were submitted by NHK (Japan Broadcasting Corp.) and 16 commercial broadcasters in the three metropolitan areas. Upon receipt of this report, MPHPT grants preliminary permits to those broadcasters on 18 April 2003 (p. 126).
	5	14	MPHPT announces outlines of the results of the survey and the assessment concerning actual radio spectrum use in fiscal year 2002 (p. 132).
	6	6	Radio Law amended (103-2: Spectrum User Fee): Cost of specific spectrum change adjusted with broadcasters (2003Law68) (p. 136).
	6	25	MPHPT receives partial report concerning effective use of mobile commercial-use frequencies in the 800 MHz range (p. 125).
	6	27	MPHPT invites comments on the draft outline of the second report from Study Group on Policies Concerning Effective Radio Spectrum Use (p. 133).
	7	30	MPHPT inquires of the Telecommunications Council about the ‘mid- to long-term outlook of radio spectrum use and the role to be played by the administration’ and the Council submits a report (‘Radio Policy Vision’) (p. 131).
	7	30	MPHPT announces reports concerning the calculation of financial compensation for the reallocation of the radio spectrum (p. 133).

Year	*Month*	*Day*	*Event*
	7	30	MPHPT receives a report from the Telecommunications Council about Radio Policy Vision (p. 131).
	9	30	Second report of the 'Study Group on Policies Concerning the Effective Radio Spectrum Use' announced (p. 133).
	10	10	MPHPT develops the 'Guidelines for Radio Spectrum Reallocation' (p. 129).
	12	1	Terrestrial DTV service started in certain districts of each of the three major metropolitan areas (Tokyo, Nagoya, Osaka) (p. 126).
	12	25	Third Report released from the 'Study Group on Policies for Effective Radio Spectrum Use' (p. 134).
2004	2	2	An Interim Report of UWB Radio Systems Committee compiled.
	3	5	MPHPT announces results of an invitation to comment on issues towards the revision of Spectrum User Fee System compiled by the Study Group on Policies for Effective Radio Spectrum Use (p. 137).
	3	17	MPHPT announces results of the survey of actual radio spectrum use in fiscal 2003 as well as the outline of the evaluation results (p. 132).
	5	19	Radio Law amended (25: Publication of Information Concerning Radio Stations): Information concerning radio stations created by registration to be disclosed (2004Law47Article2) (p. 131).
	5	19	Radio Law amended (27-18 to 27-34: Spectrum Use by Registration): Creation of a system for opening a station by registration (in place of licensing) (2004Law47Article2) (p. 123).
	5	19	Radio Law amended (71-2: Specific Frequency Change and Frequency – Termination Support Service): Creation of a system for frequency termination for specific purposes (2004Law47Article1and2) (p. 133).
	5	19	Radio Law amended (76-2-2: Restriction of Registration): Give MIC authority to restrict opening a station by registration in case of congestion (2004Law47Article2) (p. 123).
	5	19	Radio Law amended (76-3: Termination of Licence): Gave MIC authority to alter or terminate a licence on evaluation of actual radio spectrum usage (2004Law47Article1and2) (p. 132).
	5	19	Radio Law amended (103-2: Spectrum User Fee): Cost of specific spectrum termination (for reallocation) added as an objective of expense (2004Law47Article1and2) (p. 133).
	7	22	MPHPT invites comments on draft final report from Study Group on Policies Concerning the Effective Radio Spectrum Use (p. 137).
	7	22	Nikkei and Asahi Newspapers report that manufacturers of automobiles and electronic appliances expressed their disagreement with MPHPT's proposal to charge spectrum user fees with the spectrum used for electronic transportation (ETC), RFID, or wireless LAN. NC (2004.8.1, pp .82–83). Asahi (2004.7.22) (p. 137).

8	6	MPHPT invites comments on a draft plan for IMT-2000 frequency assignment in the 800 MHz frequency band (p. 125).
8	15	Softbank Broadband, Inc. (SBB) announces its intention to apply for an 800 MHz licence in order to initiate its IMT-2000 service. NC (2004.8.15, pp. 56–57) (p. 125).
8	31	MPHPT announces 'Action Plan for Radio Spectrum Reallocation' (p. 129).
9	10	Ministry changes its English name from MPHPT to MIC.
10	1	MIC announces the Final Report from the Study Group on Policies Concerning Effective Radio Spectrum Use (p. 133).
10	13	SBB files an appeal to a district court in opposing MIC's allocation plan of the 800 MHz band for IMT-2000. NCI (2004.10.1). NK (2004.10.14) (p. 125).
10	20	MIC sets up a Study Group to Examine Expanded Use of Frequency Bands for Mobile Services (SG-800 MHz) (p. 125).
10	20	First meeting of SG-800 MHz. NCI (2004.10.20) (p. 125).
11	4	An intention to start IMT-2000 service announced by the e-Access Corporation, a wireless provider. NCI (2004.11.4) (p. 125).
11	4	Second meeting of SG-800 MHz. NCI (2004.11.4) (p. 125).
11	8	MIC invites comments concerning draft concrete plans for calculating spectrum usage fees (p. 137).
11	8	Third meeting of SG-800 MHz. NCI (2004.11.8) (p. 125).
11	25	Fourth meeting of SG-800 MHz. NCI (2004.11.25) (p. 125).
11	29	MIC receives a report from the Information and Communications Council on the technical conditions for 5 GHz band wireless access systems (p. 127).
12	6	SBB approved to file with MIC an application for an 800 MHz licence for IMT-2000; SBB withdraws the petition previously filed with the district court. Asahi Newspaper (2004.12.6) (p. 125).
12	14	Fifth meeting of SG-800 MHz. SBB proposes a new plan for reallocating the 800 MHz bands. NCI (2004.12.14) (p. 125).

Source: Compiled by the author by using materials published in printed form or on the Web by the Japanese Parliament and MIC, as well as articles published in various newspapers and periodicals. (http://www.soumu.go.jp/joho_tsusin/eng/index.html, http://www.soumu.go.jp/joho_tsusin/pressrelease/joho_press.html, http://itpro.nikkeibp.co.jp/NCC/, http://www.nikkei.co.jp/, http://itpro.nikkeibp.co.jp/)

Notes
1 Abbreviations: NC: *Nikkei Communication* (Monthly, J).
2 NK: *Nikkei Shimbun* (daily newspaper, J); NCI: *Nikkei ITPRo Homepage* (web page, J). (J: Materials written in Japanese.)

6 R&D and intellectual property in a changing Japanese telecommunications market

Masako Wakui

Introduction

Research and development (R&D) in the Japanese telecommunications industry has been characterized by outstanding leadership from Nippon Telegraph and Telephone Corporation (NTT). With the technical competence, and as the biggest telecommunications service operator in Japan, NTT also has a significant influence in setting the technical standards. Several equipment manufacturers, especially four telecommunications companies called the 'NTT family', second NTT's leading position. They maintain a close relationship with NTT and often conduct joint R&D with NTT.

The R&D system is coming to a turning point in a changing business environment, with factors such as liberalization, globalization and conversion from the plain old telecommunications services (POTS) to Internet Protocol (IP) services.[1] NTT's expenditure in R&D is still higher than any other, but the focus and managing structure have been changed slightly. The market for manufacturers has become global, where many operators interact with each other, and they are now aiming at more than just obtaining NTT's content. Standardization activities have become more globalized, where Japanese companies are trying to establish influence among such numerous and various actors.

At the same time, worldwide reinforcement and emphasis in intellectual property is taking place. Telecommunications companies are intensifying their patent right portfolios and enforcement. Free, or inexpensive, licence fees were common among major telecommunications manufacturers. They did not care very much about patents as long as they could freely produce and sell products without objection from other patent holders, as they could recoup their R&D expenditure from the product. This practice has been in NTT's interest, as it has reduced the cost of procurements and equipment, which enabled NTT to attract more customers to its new services. The telecommunications companies' close relationship, led by NTT in R&D, also helped coordinate access to the intellectual property.

These basic conditions are now changing. The effort to derive profit from R&D and licensing is becoming more evident in the telecommunications industry. Being sued by a maverick patent managing company is something that Japanese companies pre-eminently want to avoid. The deadlock once caused by Qualcomm's patent enforcement announcement in the process of third-generation (3G) mobile telecommunication standard-setting is remembered as significant of the potential of a patent right to destroy the standardization effort. Microsoft's dominance resulted in its position as the single *de facto* standard-maker, which made companies fearful that worldwide-dominant companies would take exclusive possession of standards. Anxious Japanese companies and the Japanese government are trying to devise measures to counter the situation, but whether these trials will be successful remains uncertain.

What is the future for the Japanese telecommunications R&D system? Will the coordinated R&D and licensing system topped by NTT turn into a system organized in an *ad hoc*, chaotic way and dominated by a number of actors? What friction is being caused by the transformation? Will the efforts to establish some kind of order and coordination succeed?

Though these phenomena can be seen all over the world, the R&D system in the telecommunications industry takes a different shape in each country. Telecommunications services had a decisive influence on R&D structures, and have been closed within a country or region, because of the high entry barrier set by governments and technical regulations. Each country has different intellectual property rules and cultures. Competition law, which regulates licensing practices, is also different. To examine the issue, taking into consideration these backgrounds, the following section explains and analyses the issue, focusing on the situation in Japan.

R&D system

R&D under the 'Carrier's Specifications'

Nippon Telegraph and Telephone Corporation (NTT) has always been pre-eminent in telecommunications R&D in Japan. The result has contributed not only to an advanced telecommunications service and infrastructure, but also to the whole information and electronics industry in Japan. The researcher has also been active in presenting results internationally and academically.[2] NTT's human resources portfolio has been significant[3] and has attracted the best engineering and science students in Japan.[4] It appears to be able to claim higher R&D competence in telecommunications technology than any other institute in Japan, even universities and government-funded research institutes. Its importance to the Japanese economy caused the government to allow NTT to keep its R&D section at its core, NTT Holdings Company, when NTT was reorganized into the separate group companies of NTT East and NTT West (regional services), NTT Communications (long-distance services), NTT DoCoMo (mobile phones) and NTT Data. The NTT management were so keen to reserve the section as it was that they even agreed to reorganization, with which they had once strongly disagreed.[5]

The Japanese telecommunication equipment industry was established under NTT's influence and leadership.[6] It originally comprised four manufacturers – NEC, Hitachi, Fujitsu and Oki – collectively called the 'NTT family'. They were given a privileged position by NTT until the early 1980s, when the NTT procurement market was opened to foreign companies under US pressure. The manufacturers also engaged in cooperative R&D activities, sometimes only with NTT and sometimes all together.[7]

Meanwhile, other manufacturers with a different orientation, such as a part of a large conglomerate (Mitsubishi Electric), an integrated electric equipment manufacturer with wide coverage from heavy electric machinery to consumer products (Toshiba), or in consumer electronics and electrics (Matsushita), were added as partners to NTT, and foreign companies such as IBM undertook joint R&D with NTT.[8] The companies of the 'NTT family' established their business outside the telecommunications industry, where their position in these industries was more or less independent from NTT. However, the basic structure of R&D in the telecom sector stayed the same: NTT is always a leader in R&D and a limited number of manufacturers are second; these normally compete with each other and sometimes cooperate with each other, but always see the relationship with NTT as critical.

The R&D system was reinforced by, and reinforces, the so-called Carrier's Specifications practice, in which NTT takes the lead in setting the specifications of terminal equipment and facilities, and manufacturers produce them accordingly.[9] The structure was created by NTT, which does not manufacture equipment itself, but keeps several manufacturers in competition and ensures that their products improve under NTT's specification and quality control.

The same sort of structure exists in the cable and wire industry. Here, NTT has been keeping a close relationship with three Japanese companies: Sumitomo Electric Industries, Furukawa Electric and Fujikura. NTT has procured its cable and wire from these three companies, which also cooperate with NTT in R&D.[10]

NTT's R&D expenditure in fiscal year (FY) 2003 was ¥354,862 million in total. The expenditure of KDDI Corporation, the second biggest telecommunications operator in Japan – created out of the merger between Kokusai Denshin Denwa Co. Ltd (KDD) (the former statutory monopolist in international telecommunications services) with two long-distance and mobile companies, DDI Corporation and Nippon Idou Tsushin Corporation (IDO) – is about a twenty-fifth of NTT's: ¥13,339 million (FY 2003). The R&D expenditure of Softbank – a newly entered maverick company – is even smaller: ¥1,267 million in FY 2003.[11] Table 6.1 shows R&D expenditure in the telecommunications industries.[12] NTT is

Table 6.1 R&D ranking: telecommunication service, 2003–2004

	R&D investment (£m)	*R&D investment operating profit (%)*	*R&D investment sales (%)*	*Operating profit (£m)*	*Sales (£m)*
NTT	2,063.94	28.0	3.6	7,367	56,936
Deutsche Telekom	634.16	21.1	1.6	3,005	39,345
France Telecom	336.81	4.4	1.0	7,677	32,498
BT (UK)	334.00	11.8	1.8	2,822	18,519
Telstra (Australia)	250.84	10.8	2.9	2,325	8,626
TeliaSonera (Sweden)	197.43	18.5	3.1	1,068	6,399
Vodafone	171.00	n/a	0.5	−4,452	33,559
AT&T	154.74	5.3	0.8	2,936	19,289
KT (South Korea)	134.09	31.6	2.5	425	5,427
SK Telecom (South Korea)	105.13	7.3	2.4	1,437	4,463
Telecom Italy	97.94	2.4	0.5	4,036	21,738
KDDI	54.52	7.4	0.4	734	14,518
Telefonica (Spain)	51.66	2.4	0.3	2,155	20,011
Belgacom (Belgium)	44.39	12.6	1.2	352	3,789
Telenor (Norway)	38.71	5.2	0.9	742	4,441

Source: Department of Trade and Industry (UK), *The 2004 R&D Scoreboard: Company Data Part 2.*

the world leader in R&D expenditure. Although the biggest expenditure does not necessarily mean the best R&D, the figure does at least indicate that NTT is one of the world's keenest innovators.

The expenditures of manufacturers are shown in Table 6.2. The amount is the total expenditure on their various R&D activities. Japanese manufacturers have large coverage of various technology areas, and their expenditure in telecommunications-related R&D is only part of their total R&D spending, which suggests that NTT's position is even more significant. For example, Hitachi's spending in information communication was ¥169,882 million (£884.8 million) in FY 2003. Toshiba's expenditure is broadly aimed at semiconductors, digital TV and power plant as well as information communication systems. Cable and wire companies – Sumitomo Electric Industries, Furukawa Electric and Fujikura – respectively spent ¥55,276 million (¥21,600 million in telecom and electronics), ¥19,066 million (¥14,752 million in telecom and electronics) and ¥11,000 million (¥100 million in information communication and electronics) on R&D.

Loosening of the tie?

A slight change in R&D capacity balance has been noticeable recently, though the basic picture has stayed the same.

NTT's R&D expenditure has been rather stable, ¥350–400 billion, despite the worldwide telecommunications industry's downturn and the depression of the Japanese economy.[13] A closer look reveals slight changes, however. The focus has been shifting towards applied technology rather than basic research, which until now has been the pride of NTT.[14] Since the restructuring of NTT, a unified R&D effort among NTT group companies is also sometimes difficult. The gap between NTT DoCoMo, of which NTT Holdings holds only 60 per cent and which to an extent is independent from NTT Holdings' control, has become significant. DoCoMo management sometimes express dissatisfaction; for instance, DoCoMo will say that it doesn't expect much (or even anything) from NTT Holdings' research section and that DoCoMo is just contributing to its costs, as it has been commanded to do, etc. It is said that NTT is no longer a single united entity now that group companies have started competing. This might mean that mentioning the R&D expenditure of NTT as a whole might not be absolutely relevant in understanding the R&D status. Each NTT segment respectively spent the following on R&D: ¥168.9 billion (regional, or NTT West and East); ¥32.5 billion (long-distance, or NTT Communications); ¥124.5 billion (mobile, or NTT DoCoMo); ¥15.4 billion (data communication, or NTT Data); and ¥160.4 billion for other R&D activities, which include basic R&D of commonly used technologies in FY 2003.[15] Separated out, the amount looks less impressive than it did, though it is still significant.

The links among Japanese companies, especially ties with NTT, have

Table 6.2 R&D ranking: IT hardware, 2003–2004

	R&D investment (£m)	*R&D investment operating profit (%)*	*R&D investment sales (%)*	*Operating profit (£m)*	*Sales (£m)*
Nokia	2,802.99	80.6	13.5	3,479	20,755
Intel	2,435.62	59.9	14.5	4,068	16,838
Ericsson	2,275.52	n/a	24.9	−842	9,141
Motorola	2,106.59	368.3	13.9	572	15,115
Hewlett-Packard	2,040.11	122.7	5.0	1,662	40,814
Hitachi	1,938.10	199.5	4.3	972	44,996
Toshiba	1,755.09	232.9	6.0	754	29,083
Cisco Systems	1,751.30	72.0	16.6	2,432	10,546
Fujitsu	1,489.37	n/a	6.2	−707	24,069
NEC	1,337.86	184.2	5.2	726	25,576
Nortel Networks	1,245.74	n/a	21.1	−2,178	5,899
Alcatel	1,122.46	n/a	12.7	−1,177	8,817
Sun Microsystems	1,026.20	n/a	16.1	−1,551	6,387
Texas Instruments	976.48	192.5	17.8	507	5,494
Mitsubishi Electric	842.84	>999.9	4.4	59	18,968

Source: Department of Trade and Industry (UK), *The 2004 R&D Scoreboard: Company Data Part 2.*

also been gradually changed. Although joint R&D is still conducted, as seen in Next Generation Network[16] and NTT DoCoMo's 3G mobile terminal equipment,[17] the formation is increasingly becoming merit and *ad hoc* based, which is reflected in joint patent ownership practice, as detailed later.

Several factors are behind this change. One is liberalization of the telecommunications service, which brought numerous service providers into the market. Another is that the market for manufacturers has to be global in order for them to recoup their ever-increasing investment and that, outside Japan, many operators interact with each other, and NTT is only one of them. The management of the manufacturer's NTT family recognizes the need to broaden their focus by not only acquiring content solely from NTT.[18]

Standard setting

Technical regulations and standards: framework

In Japan, the Minister of Internal Affairs and Communications legislates and enforces telecommunication technical regulations. Under the Telecommunications Business Law, the telecommunications carrier installing telecommunications circuit facilities is ordered to maintain those facilities in compliance with the technical conditions specified in the applicable ministerial ordinance of the Ministry of Internal Affairs and Communications (MIC) (TBL Article 41). MIC also specifies the technical conditions for terminal equipment (TBL Article 52). The terminal equipment manufacturers complying with the conditions are given the position that the telecommunications carriers cannot refuse the request for the connection of terminal equipment that complies with specific technical conditions.[19] The details of the conditions are specified in Regulations for Telecommunications Facilities for Telecommunications Business (Ministerial Ordinance of MPT No. 30 of 1985) and Ordinance Concerning Terminal Facilities etc. (Ministerial Ordinance of MPT No. 31 of 1985).

For radio facilities and equipment, the Radio Law additionally set the requirement (RL Chapter III) and compliance with these requirements is a condition of the licence (RL Chapter II). The law particularly sets the following requirements: the quality of radio waves from transmitting equipment, the tolerance and bandwidth of frequencies and intensity of harmonics, etc., must conform to requirements specified in MIC ordinance (RL Article 28); receiving equipment shall not disturb the function of other radio equipment by its incidentally produced radiation or its high-frequency current in excess of the limits specified in MIC ordinance (RL Article 29); and the radio equipment must be installed with safety devices as specified by MIC ordinances in order to avoid harm to human bodies or damage to other objects. Ordinance Regulating Radio

Equipment (Radio Regulatory Commission Rules No. 18 of 1950) details the requirements.

The technical conditions under the Telecommunications Business Law are to be set after inquiry of the Telecommunications Council (TBL Article 169). The Radio Law requires that the MIC has to consult the Radio Regulatory Council when it specifies requirements in the ordinance (RL Articles 99-11). The Telecommunications Council is composed of 30 regular committees, some special and *ad hoc* committees that are appointed by MIC. At the time of writing, three members of these committees are persons who are working for telecommunications equipment manufacturers (Hitachi, Sharp, Toshiba and Lucent), and many more (15) are academics. Under the council, an Information Communication Technology Committee is established, which takes charge of detailed study and examination of the technological issues concerning electronic communication and the use of bandwidth. The Radio Regulatory Council is composed of five members: three academics and persons working for software or insurance companies.

Telecommunications standards with which compliance is not mandatory are set by the Telecommunications Technology Committee (TTC) regarding telecommunications networks and terminal equipment and the Association for Radio Industry Businesses (ARIB) in regard to utilization of radio waves. Though they are private standard-setting organizations (SSOs), composed of and financed by telecommunications service operators and manufacturers, they are recognized by MIC as organizations to develop Japanese standards that conform to International Telecommunication Union – Telecommunication Standardization Sector (ITU-T) recommendations. They are also ITU-qualified organizations for cooperation and exchange of information with ITU (ITU-T Recommendation A.6). Because of the need for interconnectivity and interoperability, their standards are basic documents to be referred to by companies that have any interest in the telecommunications service or manufacturing businesses in Japan, even though their standards are not mandatory. The standards or any works conducted by the members, or discussions in these organizations, are naturally often considered by MIC. The standard occasionally becomes mandatory through reference in the MIC ordinance as well.

NTT as the standard-setter

NTT has been taking a leading role in standard-setting activity in Japan. Along with NTT's position as the biggest telecommunications service provider and controller of networks, its expertise and competence in technologies have secured NTT significant influence in standard-setting for NTT. Though formal and regular members of the MIC council are supposed to be independent and neutral, and it does indeed consist mainly of academics, when technical details needed to be examined the knowledge and expertise of the industry are summoned. The experts have regularly been from NTT or manufacturers who conducted R&D with NTT. NTT has been also a major member of the TTC, ARIB and the consortium that was formed to set the *de facto* standards.[20]

NTT's leading position has derived from its outstanding commitment in international standard-setting organizations as well. Along with KDDI, the successor of KDD, which previously had a monopoly in international telecommunications services, NTT has been constantly submitting contributions, as many as half the total contributions from Japanese members,[21] sending its specialists to international standard-setting organizations, particularly ITU since the 1950s[22] and chairing the ITU subcommittee.[23]

Towards chaos?

Though NTT's leading position still seems to be solid, standards that are alien to, or set independently from, NTT are becoming significant in the telecom industry these days. Yahoo! BB, which has a larger market share than either NTT East or NTT West, adopted and implemented Annex-A protocol for its broadband service, which is different from Annex-C, used by other broadband service operators, including NTT. In the mobile market, DDI (a competitor of DoCoMo, now merged into KDDI) adopted the CDMA (Code Division Multiple Access) standard supported by Motorola, in spite of the existence of the PDC (Personal Digital Cellular) standard, which was nationally promoted. For 3G, KDDI adopted cdma2000 in cooperation with Qualcomm, after cdmaOne for 2.5 generation, while NTT is promoting W-CDMA. As for the Internet, most of the standards are set outside the traditional formal organizations, such as the Internet Engineering Task Force (IETF), World Wide Web Consortium (W3C) or Institute of Electrical and Electronics Engineers (IEEE), where Japanese companies and institutes have less influence compared with the traditional formal standard-setting bodies[24] in spite of their attentions.[25] In these standard-setting bodies, the nature and size of the companies involved are more various, and the people less acquainted with the 'nation's pride' company having played major roles in telecommunications standard-setting.

The recent non-NTT standard implementations are rather different and are more significant in size and method. From time to time, NTT and major manufacturers have implemented the advanced technology and standardized their systems accordingly. They often adopt international standards to get better manufacturing competence abroad. Motorola brought its standard JSMR (Japan Shared Mobile Radio) system to Japan in 1986 in competition with MCA (multi-channel access), which was supported by Japanese companies. DDI Corporation, in cooperation with Motorola, adopted TACS (Total Access Communication System), which was widely used in Europe but new to Japan in the 1980s. These are examples of the introduction of foreign-made standards. But in these cases, the adoption was either by NTT or established manufacturers, or at least was done in a more organized way. In the case of JSMR and TACS, the market share was not as significant as in recent cases.

Recent non-NTT standards were implemented more competitively and on a larger scale, and the government did not take a notable role in the implementation. While Motorola had to use pressure by the US government to nudge the Japanese government into a position to be able to promote its supporting standard,[26] Yahoo! BB spread its service first and went to ITU to present the problem of the Japanese standard-setting system against the will of TTC. The government was involved only after the difference in the standard caused serious business dispute.[27] As for the Internet, the Internet network in Japan was first set up by academics at JUNET (Japan University Network), then by WIDE (Widely Integrated Distributed Environment), which the Ministry of Posts and Telecommunications – the telecom regulator at that time – had less influence on.

One factor driving the change is globalization. The overall globalization of economic activities necessitates a seamless network, for which international interoperability and connectivity are vital. Telecommunications companies need a wider market with more customers to recoup the ever-increasing R&D costs, too. To implement internationally recognized standards is essential in materializing their global competence.[28]

Another factor is liberalization. The competitive telecommunications operators recognize that they are at a disadvantage to NTT under the 'NTT specifications' and R&D system. Their disadvantage stems from their inability to access the advanced equipment and facilities, exemplified by the administrative warning case at the Japan Fair Trade Commission.[29] In this case, NTT DoCoMo was warned against its licensing practice of permitting mobile terminal equipment manufacturers to produce and sell terminals for competitors' service on DoCoMo's patent licence only a few months after the terminal equipment for DoCoMo's service had been introduced onto the market. Japan has neither standardized mobile terminal specifications nor a SIM card system, and specifications for the mobile terminal equipment are slightly different for each network.[30] Against this background, DoCoMo's practice effectively disadvantaged NTT competitors by denying their subscribers the advanced equipment. Even without such obvious restrictive conduct, this kind of disadvantage will remain as long as there are insufficient facilities and equipment free from the influence of NTT. The vulnerable position of competitors, as well as the need to differentiate their services from those of NTT, has been driving them to adopt standards established abroad.[31]

Patent ownership and licensing

Harmonized use of advanced technology

Changes in the R&D and standardization system in the telecommunication sector in Japan are changing patent practices accordingly.

Figure 6.1 shows the number of NTT patent applications made with

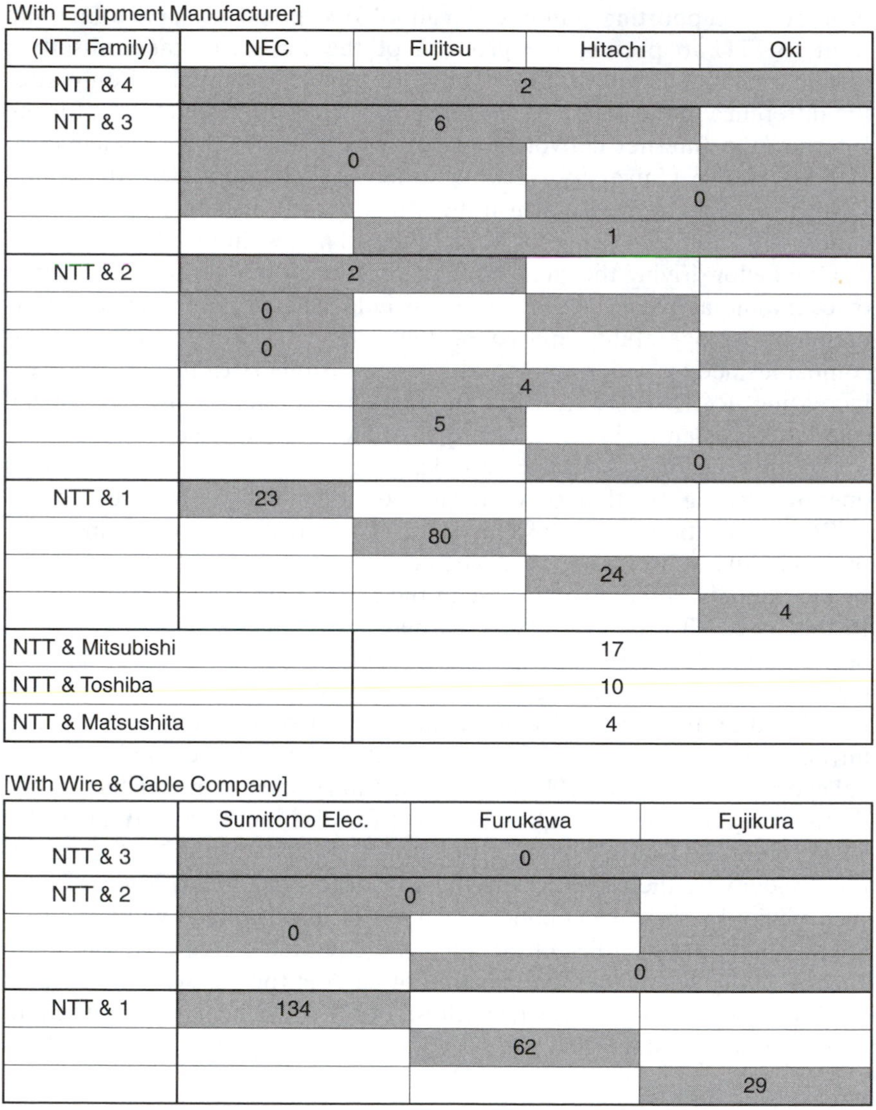

Note: (NTT application total: 3343)

[With Equipment Manufacturer]

(NTT Family)	NEC	Fujitsu	Hitachi	Oki	Applications
NTT & 4	■	■	■	■	0
NTT & 3	■	■	■		0
	■	■		■	0
	■		■	■	0
		■	■	■	0
NTT & 2	■	■			0
	■		■		0
	■			■	1
		■	■		2
		■		■	2
			■	■	6
NTT & 1	■				12
		■			12
			■		10
				■	11
NTT & Mitsubishi					4
NTT & Toshiba					0
NTT & Matsushita					7

[With Wire & Cable Company]

	Sumitomo Elec.	Furukawa	Fujikura	Applications
NTT & 3	■	■	■	0
NTT & 2	■	■		0
	■		■	2
		■	■	0
NTT & 1	■			4
		■		9
			■	1

(NTT application total: 2549 (DoCoMo not incl.))*

[NTT DoCoMo & Equipment Company]
NEC 3; Fujitsu 4; Hitachi Kokusai 8; Oki 0; Mitsubishi 4; Toshiba 0; Matsushita 1
(NTT DoCoMo application total: 618)

Figure 6.1 Joint patent applications with NTT in 1993 and 2003 (source: Author's analysis based on Patent Office Gazette Database).

Note
All patent applications from NTT DoCoMo (called NTT Mobile Communications Network, Inc., at the time) were with NTT.

major Japanese manufacturers and published in the years 1993 and 2003.[32] It shows a significant number of joint applications in 1993. The practice of joint patent ownership has been thought to promote smooth implementation of newly developed technologies.[33] Under Japanese patent law, jointly owned patents cannot be licensed without both parties' consent, though they can be practised independently. This means that companies normally apply for patents jointly when they are in a steady relationship and they are sure that their licensing policies in the future will be aligned with each other. The joint patent application accordingly reflects close ties between applicants, which can be observed in Figure 6.1a.

As for patents owned by NTT alone, NTT has been generous in enforcing its large portfolio against manufacturers. In cases where the standard is involved, NTT has been especially willing to grant a licence easily.[34] Figure 6.2 shows the number of patent licensing declarations submitted to ITU-T by major telecommunications operators and manufacturers. ITU has a patent policy stating that provided the patent owner submits a patent statement and licensing declaration, it will be willing to license the patent for free (so-called 'option 1') or on non-discriminatory and reasonable terms ('option 2'). The licence may be conditional on the prospective licensee committing itself to licensing its essential patents for implementation of the same for free or under reasonable terms and conditions ('reciprocity'; in the figure, 'RPCC'), which is also stated in the licensing declaration.[35] As seen in Figure 6.2, NTT has been quick to submit the licensing declaration and has occasionally chosen option 1. The generosity has been easing the patent burden on manufacturers as well as joint patent ownership practice.

Securing access to the new technology without onerous patent licensing negotiation and fees has been in NTT's interest. NTT understands that setting high royalty rates is not feasible if it wants to promote its new service, which is its main source of revenue, and also to reduce its procurement costs. NTT has accordingly adopted a basic policy towards licensing of setting reasonable royalty rates.[36] Though NTT is now emphasizing the revenue from the patent, and some changes have been witnessed by the industry, a significant alteration in its position is still to be seen.[37]

Among the major manufacturers, implicit cross-licencing or non-enforcement has been a common practice,[38] though they are basically keener to enforce their patents than are service operators. Cross-licencing and implicit licensing are natural results of the limited number of patentees, each having a significant patent portfolio of like size steadily needing the patent licensing of others. In the telecommunications market, the patent joint ownership practice and the influence of NTT, and the government, stabilized the practice further. NTT was not in favour of expensive royalties that would result in higher procurement costs and terminal equipment prices. The government shared the view of NTT that the newly developed technology should be spread as rapidly as possible.[39]

In general, the most feared patentees are the independent R&D companies, or the specialized patent licensing companies that do not manufacture products. To them, the patent portfolio of established companies cannot be used as a 'mutually assured destruction' weapon,[40] or a threat countering their patent enforcements, as they do not have anything to be threatened by others' patent enforcement. In the Japanese telecommunications market, however, such companies have not been successful so far.[41]

Surviving pro-patent in the diversified world

A change in this practice to a stronger position in enforcing patents can now be seen. The emphasis on intellectual property is a general trend in Japan, and in this respect the telecommunications sector is no exception. Their investors are starting to demand more explicit results from the company's investment in R&D.[42] Like other advanced countries, Japan sees intellectual

	ITU-T	*83&84*	*85–89*	*90–94*	*95–99*	*00–04*	*[General]*	*ITU-R*
NTT	Option 1	1			6	2		0 + 2
+	(RPCC)							
NTT	Option 2			1	64 + 3	1 + 1		
DoCoMo	Option 2							
	(RPCC)		1	4		2 + 1		
KDDI	Option 1							
	(RPCC)					1		
(KDD –								
Oct 2000)	Option 2			6	5			
	Option 2							
	(RPCC)		2	1		20		
Fujitsu	Option 2			2	5	2	1('99)	80
	Option 2							
	(RPCC)		1		38	15		
NEC	Option 1							
	(RPCC)	1			5	2		5
	Option 2							
	(RPCC)		2	2	11	1		
Hitachi	Option 2		1	1				
	Option 2							
	(RPCC)				7			
	Option 1							
	(RPCC)				6	8		1
Mitsubishi	Option 2			2	4			
	Option 2							
	(RPCC)		1	3	15	9		
Oki	Option 2		1	1	2			48

	ITU-T	*83&84*	*85–89*	*90–94*	*95–99*	*00–04*	*[General]*	*ITU-R*
France Telecom	Option 1 (RPCC)	1						
	Option 2			1	6		1('03)	
	Option 2 (RPCC)		1			8		
Alcatel	Option 2		1		9		1('99)	
	Option 2 (RPCC)				6	9		
BT	Option 1 (RPCC)	1						
	Option 2		1	1	4			
	Option 2 (RPCC)		2	1	2	8		
Deutsche Telecom	Option 2 (RPCC)					2		
Siemens	Option 2			6	7			1
	Option 2 (RPCC)				8	13		
Robert Bosch	Option 2				5		1('02)	
	Option 2 (RPCC)					16		
KPN	Option 2				1			
	Option 2 (RPCC)				2	24		
Nokia	Option 1 (RPCC)			5			1('01)	
	Option 2 (RPCC)				3	9		
	Option 1 (RPCC)					1		44
Ericsson	Option 2				3			
	Option 2 (RPCC)				2	2		
Nortel	Option 1 (RPCC)	1			1		1('99)	
	Option 2 (RPCC)		1*	10	1	8	10	
AT&T	Option 2				4			
	Option 2 (RPCC)						5	
WorldCom	Option 2 (RPCC)					24	1('00)	
Lucent	Option 2				3		1('99)	1
	Option 2 (RPCC)			11	21	7		
IBM	Option 1 (RPCC)					23*	Note: The declarations were submitted for T.88 &T800	
	Option 2		1	5	2			

	Option 2 (RPCC)	1*	20	6	17		
Intel	Option 2 (RPCC)			24	21		
Motorola	Option 2		2	4			11
	Option 2 (RPCC)			5	5		
Cisco Systems	Option 2			6		1('03)	
	Option 2 (RPCC)				14		
Texas Instruments	Option2		2	16	8		
WorldCom	Option 2 (RPCC)				24	1('00)	
Lucent	Option 2			3		1('99)	1
	Option 2 (RPCC)		11	21	7		

*year unknown

Figure 6.2 International Telecommunication Union (ITU) patent licensing declarations (source: Author's analysis based on ITU Patent Database).

property as the key to Japanese industry staying vital. The Japanese government regards prosperity as one of its top policy objectives, and has implemented several measures, such as anti-counterfeit programmes in Asian countries; setting up the Intellectual Property High Court, which specializes in intellectual property matters and is expected to help bring speedy and effective enforcement; the patent licensing adviser and agents scheme; and so on.[43] The basic idea is that Japan needs to transform itself from a manufacturing-based economy to a knowledge- and innovation-based one, against its loss of comparative advantage to Asian countries, especially China, where labour costs are much lower than in Japan.

Against this background, the 'mutually assured destruction' weapon is becoming less effective as the manufacturer's interest shifts away from revenue from manufacturing to that from intellectual property. Whether Japanese companies like patent activism or not, they cannot avoid the patent activism of foreign companies anyway. Internationally, a material number of ventures have appeared and started establishing their positions on the basis of patent portfolios. The mitigating factor described above does not work as well as in Japan.

Patents that have to be licensed to comply with a standard are under focus, as these have become a field where Japanese companies cannot escape from tough negotiations. The idea of claiming and enforcing the essential patents covering standards was new, especially to the telecommunications sector, where it was considered that standards should be accessible by all. Moreover, standards were seldom covered by a patent when they did not need to set out a detailed description of the technologies, nor were standards written based on the latest technologies. To the extent that standards were under NTT's influence or cross-licensed, the potential licensees had little need to be bothered anyway.

Now the environment has changed notably.[44] International standard-setting has become more important, and the various companies have started to ensure that their intellectual property rights (IPRs) do not lose their value simply because they are implemented within the standard. The revenue from licensing the patent covering internationally adopted standards is widely recognized, and the stake in the patent has become too high to lose just to promote a standard. Hearing the news that Qualcomm intended to set a high royalty rate for the 3G mobile standard shocked Japanese companies. This news was followed by the nightmare of Ericsson likewise claiming high royalties, and the standard-setting activity was put on hold for a while. Now, however, Japanese manufacturers too do not hesitate to show their interest in gaining income from their intellectual properties, even to the extent of proclaiming an uncompromising stance against standard-setting.[45] Mitigating factors such as the influence of the telecommunications operator and the government are constantly losing effectiveness.

Figure 6.3 shows the presence of foreign companies in the 3G mobile

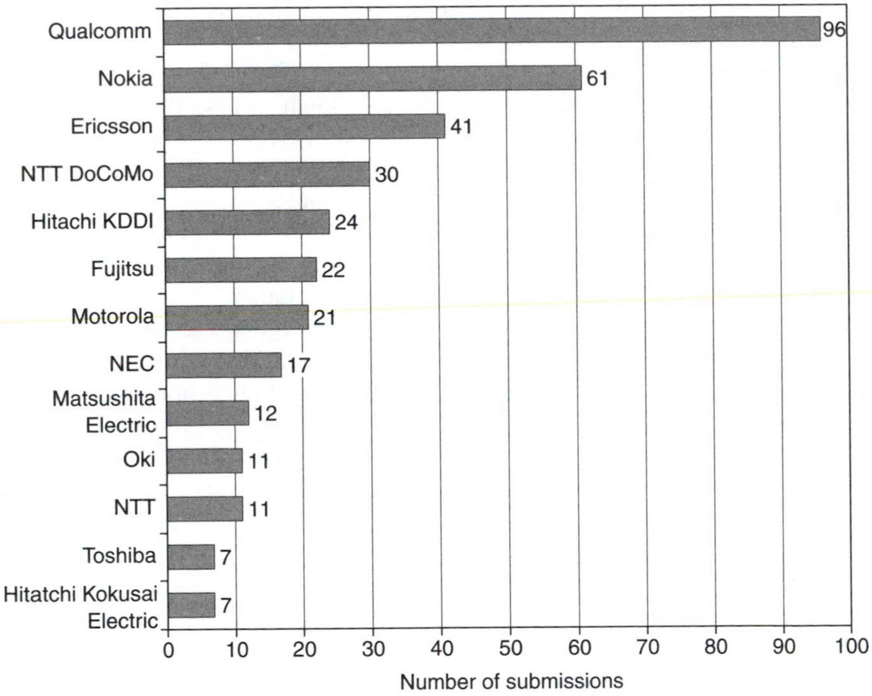

Figure 6.3 ARIB standards (W-CDMA and CDMA2000) and patents: the number of licensing statements submitted to ARIB (source: Japan Patent Office, Study Report on the Trend of Patent Application: Mobile Communication System (2003), page 17).

communication system standards patent.[46] The figure exemplifies the internationalized patent licensing environment when mobile communications standards are involved.[47]

A change in joint patent application practice can be observed as well. As Figure 6.1 shows, joint applications are still made,[48] but their number has significantly decreased, which is considered to reflect the R&D system and relationship among companies in telecommunicationsa sectors.

The real effect and seriousness of the patent thicket caused by various numerous patent holders' patent activism have yet to be seen. NTT and the established companies are still modest in their behaviour towards each other, and a real, material threat from independent technology-based companies does not exist in Japan. Companies may ultimately recreate similar informal cross-licence practices on a wider scale. Whether or not the reality changes, one thing is clear, however: Japanese companies and the government are taking the situation seriously and seeking a solution. Some options under consideration for this will be detailed in the next section.

The response: IPR policy, patent pools, antimonopoly law and compulsory licensing

Japanese companies see intellectual property as an essential source of survival in globalized competitive markets. At the same time, they also fear that patents assault their manufacturing, sales and R&D activities, feeling uncomfortable that they have to go through licensing negotiation with ever-increasing and varied patent owners. In particular, patents claimed to be essential to comply with standards are causing serious concern.

This mixed attitude is also true of the government. The Japanese government is aiming to change the nation's attitude, to turn it towards beconing an intellectual property-based country, and has implemented measures to that end. But it is also concerned about excessive patent activism.

The answer to these seemingly conflicting needs is to secure for inventors fair and reasonable revenue for inventions through intellectual property, while making sure that the property is to be accessed by a company willing to share reasonable R&D costs.

In the event that industry standards are involved, or there is a 'patent thicket' situation, or a need for lengthy licensing negotiations with numerous patent holders to implement the technology, two inefficiency concerns can be involved. One is the 'hold-up' problem, which describes a situation in which patent-holders exercise their patents aggressively and demand expensive royalties after the licensee has become locked into a standard for which patent licensing is essential. The concern can inhibit implementation of standards by those who want to avoid a 'hold-up' situation. Another is the 'Cournot' problem, or the double marginalization problem, which is caused by holders of essential patents setting their profit-maximizing prices without taking into account their royalty's negative effect on others' (and total) royalty revenue. In the end, the resulting accumulated royalty rate is too high, not only for licensees, but also in terms of patentees' total profit. Securing access to essential patents at reasonable terms can relax these concerns and promote the spread of standards and the setting activity.[49]

On the understanding that some institutional measures are considered necessary to realize the desired status, the cabinet's policy statement ordered the relevant ministries and departments to devise a measure to promote smooth use of the patent right, especially in the context of the industry standard.[50] The measures under discussion include regulations under antimonopoly law and compulsory licensing under patent law, patent pools and IPR policies of standard-setting organizations, which are detailed in the next subsection.

IPR policies of standard-setting organizations

Major SSOs have policies on how to handle patents that cover technical standards. In the telecommunications sector, the established major SSOs have similar IPRs policies, either internationally or domestically.[51] The basic framework implemented in the major standard organizations such as TTC and ARIB, as well as ITU, is as follows.[52] The party involved is asked to notify the organization once it knows it will need to infringe a patent or application that is in order to implement a standard that has been set. The owner of the essential patent is told to submit a statement either that it will licene the IPRs for free to any person, or will license them at reasonable terms and non-discriminately. The licence is called a RAND licence. The statement will be recorded and published with the standard. If the patent owner does not choose either option, the draft standard will be changed so that the patents should be avoidable. In this case, after the publication of the standard, once the SSO finds that the patent is not licensed on a RAND basis, it will consider changing the standard. In an extreme case, it will even withdraw the standard altogether.

The current IPRs policy has its limits:

1 It does not oblige the company involved to conduct a patent search; the obligation is fulfilled if it informs of IPRs' bona fide basis when it becomes aware of it.
2 The SSO does not conduct any patent research itself, and does not ensure the correctness of the information and statements that are given.
3 It is up to IPRs' owners whether they agree to a RAND licence.
4 There is no clear definition of RAND and there is no monitoring and enforcement mechanism of a RAND licence.
5 The outsider will always remain, however powerful the IPRs policy

> becomes, and an SSO does not have any means to compel it to co-operate to effectuate the IPRs scheme. The only prescribed solution when RAND conditions are broken is to change, or waive, the standard in the extreme case, which would not change the status of technology if the standard had been widely adopted within a market and switching to new standards were not practically feasible. In Japan, TTC and ARIB are indeed not involved with particular patent disputes and concerns. They have never withdrawn a standard for reasons of the enforcement of a patent right. Companies negotiate on and settle disputes without having recourse to the TTC and ARIB.

These limit the policy's effectiveness to remedy the hold-up concern. Outside the TTC and ARIB, to enhance effectiveness of the IPRs so that the hold-up concerns are relaxed, several measures have been considered such as stricter obligation for members to conduct patent searches; more thorough search of non-members' patents; SSOs keeping detailed records of discussions on patents, if any,[53] creating a database that enables users to easily discover the licensing position and also potential essential patents and applications; introduction of more detailed RAND statements that clarify royalty rates; establishment of the principle that increasing the royalty rate after a widespread standard has locked in users constitutes excessive royalties and is assumed to be not 'reasonable';[54] and promoting patent pools as an instrument to ensure RAND conditions.

To set a stronger IPRs policy is not an easy task, however. TTC adopted the IPRs policy in 1989, when many established SSOs started taking the issue seriously. Though more stringent policies might have been possible, TTC actually followed the ITU-T approach. Later, ARIB followed the ITU-R. These SSOs saw that having different IPRs policies was not feasible, considering the strong link between other SSOs, notably the ITU. The effect of any departure in IPRs policies by Japanese SSOs would be limited anyway, when an increasing number of standards are set at the international level. The most serious concern of the Japanese companies is being forced to use patents by foreign entities with wealthy IPRs portfolios. These conditions have led some Japanese entities to make an effort to create an environment favourable to them at the ITU,[55] such as proposing more active involvement of SSOs in patent searches.[56] It is not certain whether they have succeeded in creating favourable IPRs policies, persuading foreign companies that have developed their patent practice in different R&D and IPRs systems.

For IPRs policy to be in the public interest, it has to clear the anticompetitive risk as well. Companies setting standards are competitors, which are supposed to compete actively in the telecommunications service market and equipment market. Discussions among them have to be strictly limited to the technical matter, and discussion on the particular licensing term between competing patent owners or between potential licensees is not generally allowed. Discussion should never be about the price, output and future prospect of the equipment and/or service where the standards are used. This rule is necessary to keep the market competitive, but also makes it difficult to clarify the meaning of RAND and monitor the actual licensing conditions. Those who see a low royalty rate as being in their interest might press patent owners to set an unfairly low royalty rate, which is another anticompetitive concern.

Any stronger IPRs policy scheme also needs to sort out practical problems. Who will pay the cost of assessing the validities and essentiality of the patents? The anticompetitive risk suggests the desirability of an independent and competent institute to deal with licensing terms and patent analysis on patentability and essentiality.[57] The feasibility of setting up such an expensive organization has yet to be seen.[58]

Patent pools

Another idea, which is noted in many policy documents, is the patent pool.[59] Collecting patents into a pool saves negotiating time, and avoids the risk of being blocked by the patent. Compared with cross-licensing and joint control of the patent through joint ownership, as described, which has already been seen in Japan, the pool under discussion would be more formalized, with wider coverage of patents extending globally and more accessibly.

A type of patent pooling scheme was indeed formed for 3G mobile communication systems patents. Inspired by the success of the MPEG-2 patent pool, 19 European, Korean and Japanese companies, including electronics manufacturers and telecommunication service providers, produced a platform scheme that would collect the essential patents covering 3G standards. Under the scheme, patent owners and potential licensees concluded a framework agreement under which licensees could obtain licences, whether under the standard licence terms or as an individually negotiated agreement. Under the standard licence, the royalty rate was set at 0.1 per cent per patent for the first phase, and the maximum aggregate royalty to participants of 5 per cent of sales price, which applied to the agreement between each participant patent owner and licensee in a non-discriminatory manner. The licensee can obtain a licence from each individual patent owner by benefiting from a temporary licence under a pre-set term and dispute resolution measure, in the event of failure to reach an agreement. Participants are obliged to submit all essential patents to the platform. Those who want to participate can do so. The evaluation panel was set up, which constituted an independent organization to collect only patents essential to implement any 3G standards. For management, an independent patent platform company was established.[60]

After MPEG and 3G, however, patent pools have not been formed as often as expected, a situation that is unsatisfactory to those who see its great potential to promote competition and efficiency, and to help the smooth adoption and implementation of standards, especially through relaxing the concern for hold-ups and the accumulated royalty problem. The Japanese government started a study of the patent pool to eliminate any obstacles. In particular, the Japan Fair Trade Commission (JFTC) issued guidance that states how JFTC applies the antimonopoly law to the patent pool case (28 June 2005).

Patent pools have been regulated under antimonopoly law, as they involve serious anticompetitive risk.[61] If patent owners get together and pool their competing patents, they can set a monopoly price, free from the competition that was supposed to exist. Such a patent pool would have the power to exclude the competitors who want to implement technologies covered by a patent, by denying them access to the pooled patent. The patent owners participating in the same patent pool might exchange information about the product and service market in which they compete, restricting the competition. This suggests the need for regulation, but there is also a concern that stringent regulation and obscure antimonopoly rules may block the formation of efficient patent pools. Under the new guideline, patent pools are to be analysed from several points, especially whether pooled patents are valid and not competing with each other; whether anticompetitive tying of dominant patents and other patents is involved; whether measures are in place to prevent information about the patented products being exchanged; whether the patented products can be manufactured and sold free from any restrictions under the patents; whether a company outside the patent pool is granted access to the pooled patents indiscriminately and on reasonable terms; and whether potential licensees are able to negotiate on an individual basis without contracting for the whole licence of the pooled patents.

Some critics are further discussing the possibility of such measures as publishing antitrust guidance on the pre-pool stage, or patent pool formation activity, which should clarify what information exchanges and commitments competitors are allowed to make at the stage of planning and formation of patent pools; and introduction of the new option of patent statements under SSOs' IPRs policy, which would state that the company will agree to participate in a patent pool once formed.[62] Pro-patent pool critics once argued for clearing the SSO's policy to oblige members to participate in the patent pool once it is formed as a condition of membership and/or application of antimonopoly law against the party that rejects the efficiency-promoting patent pools, although the idea was not formally recognized under the new guide.

It is noteworthy that the pro-pools discussions are not always supported by the industry. Japanese companies' attitudes could be summarized by saying that they like the idea in general and they indeed want a pool to exist either when they do not have a strong enough patent portfolio or when the interest in the prompt spread of the technology is sufficiently large, but are unwilling to participate otherwise. The company that finds a good business opportunity to earn from its patents generally wants to remain outside. Absence of mutual trust, too, inhibits the formation of pools: patent owners' concern is that others may devalue their patents to reduce the allocation to the owner; that others might set the royalty too low because of their interest in promoting new services; that the patent pool organization might not enforce the pooled patents as well as it is supposed to do, etc. Some observers have expressed concern about the big and successful patent pool organizations beginning to demand excessive royalties.[63]

Antimonopoly law and compulsory licensing

Under the Japanese Patent Law, patent licensing can be compelled when practice of the patents is 'particularly necessary' in terms of 'public interest' (Patent Law Article 93).[64] Although the article has never been used, the use of a compulsory licensing under patent law was recently discussed against the alleged imminent needs to secure access to standards. The discussion on the use of compulsory licensing has not been in favour of it so far, however. The Working Group of the Ministry of Economy, Trade and Industry (METI) concluded that further consideration is needed that takes into consideration the need to strike a balance among the position and interest of SSOs, those who are implementing standards and so-called outsiders who are not willing to license their essential patents in a cooperative manner.[65] About the issue, too, various positions are observed according to various bodies' stakes in standards and patents.[66] During discussion, the idea is opposed because of the concern that patent rights will be weakened through effectuating the never-used compulsory licensing scheme as well.

Another possibility under discussion is the use of antimonopoly law to hold the refusal to license or the attempt to impose excessive patent royalties to be a violation of the law that ensures access to the essential patents.[67] Antimonopoly law prohibits the exclusion of the business activities of other entrepreneurs, thereby regarding a substantial restraint of competition in any particular field of trade as 'private monopolization' (Antimonopoly Act, Section 3). Exclusions are also regulated by the prohibition of 'unfair trade practices' (AMA, Section 19). 'Unfair trade practices' are defined by the JFTC, and include unjust refusal to deal (General Designation, Article 2), unjust discriminatory pricing (AMA, Article 3) and discriminatory treatment on transaction terms (AMA, Article 4). Such conduct is considered 'unjust' when it tends to impede fair competition. These designations are mainly aimed at preventing lessening of free competition. Reducing transaction opportunities for a competitor without proper justification has been considered to be impeding fair competition. Conduct 'dealing with another party by unjust use of one's bargaining position' could also be 'unfair trade practices' (AMA, Section 2(9)), and

JFTC designated the 'abuse of a dominant bargaining position', which includes conduct setting transaction terms in a way that is disadvantageous to the said party and unjustly in the light of normal business practices by making use of one's dominant bargaining position over the other party (General Designation Article 14).

There has been no published case that alleges unilateral patent enforcement to constitute illegal conduct under the Antimonopoly Act. A study group set up by the JFTC once published a report about the setting and usage of technical standards.[68] It covers *de facto* standard technology controlled by a single firm, a standard formulated by a group of firms, and the *de jure* standard set by public SSOs. The report states that the refusal to reveal interface information can be a private monopolization or an unfair trade practice, even if it is covered by intellectual property, especially when a firm that dominates one product market and establishes its technology as the prevailing standard declines to reveal the information and thereby excludes others from the market for complementary products. In the case of joint standard-setting, the report states that participants and outsiders who work the group to incorporate their technology into the standard and knowingly conceal their patents can violate antimonopoly law by refusing to license such patents later. Charging a high access fee is considered to constitute a violation when it amounts to illegal refusal to deal. It might be also regarded as an abuse of a dominant bargaining position when participants of joint standard-setting who did not make known their rights first despite of knowing that standards are covered by their IPRs enforce their IPRs against a firm caught after having invested in the facilities. The appropriateness of such applications is still under discussion within and outside JFTC. Recently, the possibility to meet a refusal to cooperate with patent pools by refusing a licence is also claimed to be worth considering.

Conclusion

The R&D in the telecommunications sector in Japan and intellectual property ownership and licensing practice as a result of R&D are going through a transformation brought about by changes in markets and technology. This change presents an issue to companies and the government as to what should be done. No resolution has been found to date that can either convince many people, is effective, or looks desirable as a measure.

These issues that are happening in Japan are seen in other countries too, such as the European countries, to a greater or lesser extent. Issues concerning patent pools and IPRs policies are being discussed in Europe and the United States. Each country seems to be reacting in a different way according to differences in existing research establishments, and views on intellectual property and company relationships. Further research is needed to find a solution at a global level, taking into account the current structure of the R&D and IPRs system, which have been historically shaped in a different way in different countries.

Notes

1 Against this background, the government conducted a comprehensive study on the current status of R&D in the telecommunications sector. See Telecommunications Council, Ministry of Internal Affairs and Communications [hereafter MIC], Council Report on the R&D and Standardization Strategy towards Stronger Technology Competitiveness in the Information and Communication Sector (26 March 2003) [in Japanese] [hereafter MIC Council R&D Report].

2 NTT R&D activity has resulted in 14 publications in *Nature* from 1997 to 2004, for example (Nature publishing group, 'Ranking: the number of articles published in *Nature*', http://www.natureasia.com/japan/nature/top10/200-.php. According to the Survey of Research and Development conducted by the Statistics Bureau of MIC, private companies spent most of the R&D cost, about ¥11.6 billion on communication information technology in 2002, while non-profit institutions and public organizations spent ¥1.8 billion and universities and colleges ¥3.3 billion.

3 NTT Holding, which is responsible for basic R&D and spent ¥188,500 million on R&D, has three thousand staff, a fifth of whom hold doctorates as of March 2003. NTT Co., *NTT Research and Development: 2004 Review of Activities*.

4 Kazunori Ishiguro, *IT Senryaku no Ho to Gizyutu* (The Law of IT Strategy and Technologies) (Tokyo: Sinzansya Publishing Co. Ltd, 2003) and its digest, *Kokusai Kyosoryoku ni okeru Gizyutu no Siten* (World-Leading R&D Activities of NTT as the Real Key to Japan's International Competitiveness) (Tokyo: NTT Publishing Co., 2003), detail the achievements of NTT's R&D activities. Engineers who have moved from NTT to other companies have been helping them gain technological competence. As for the situation in the early days, see Hirotsugu Shimoda, *Tusin Kakumei to Den Den Kosya* (Communications Revolution and NTT Public Co.) (Tokyo: Mainichi Newspaper Co. Ltd, 1981, pp. 130–132).

5 Junichiro Miyazu, *NTT Kaikaku* (NTT Restructuring) (Tokyo: NTT Publishing Co. Ltd, 2003, p. 123).

6 Martin Fransman, 'The Future of NEC', in *Japan's Computer and Communications Industry* (Oxford: Oxford University Press, 1995) [hereafter Fransman, *Japan's C&C Industry*], provides insight into the history of telecommunications equipment-manufacturing industry.

7 Ibid., at pp. 62, 73–79 and 121–123; Study Group, Ministry of Posts and Telecommunications, Study Report on R&D in Info-communications Industries for the Multimedia Age [in Japanese], p. 50 (1995). See also NTT Public Co., *Denki Tusin Zisyu Gizyutu Kaihatusi: Kokan hen* (The History of Domestically Developed Telecommunication Technologies: Switch), §3 (crossbar) and §5 (DEX) (Tokyo: Telecommunications Association, 1976).

8 See, for example, Report of Study Group, Ministry of Posts and Telecommunications, Study on R&D in Info-communications Industries for the Multimedia Age [Japanese], p. 43 (1995). About its background, see Yasuzou Nakagawa, *NTT Gizyutu Suimyaku* (NTT Technology Water Vein) (Tokyo: Toyo Keizai Inc., 1990, pp. 202, 208, 210) [on NTT R&D open-door policy]. Siina Takeo, *Gaishi to Ikiru: IBM tono Hanseiki* (Living with Foreign Affiliated: Half a Century at IBM) (Tokyo: Nihon Keizai Shimbun, 2001, p. 110) explains IBM's experience at that time.

9 See Japan External Trade Organization (JETRO), *Survey on Actual Conditions Regarding Access to Japan: Telecommunications Equipment*, §II (June 1998).

10 See Fransman, 'The Evolution of the Japanese Optical Fibre Industry', in *Japan's C&C Industry*. Some joint R&D cases in the 1960s and 1970s are described in NTT Public Co., *Denki Tusin Zisyu Gizyutu Kaihatusi Senrohen* (Wire) (Tokyo: Telecommunications Association, 1979, pp. 250–232, 256 and 543).

11 The total R&D expenditure of telecommunications service providers in Japan is calculated at ¥1,457,460 million (£7,590) (FY 2002). MIC, The Basic Survey on the Communications Industry (7 April 2004).

12 The exchange rates used are euro: 1.42; Japan: 192 (yen); Sweden: 12.88 (Swedish kronor) and United States: $1.79.

13 NTT's R&D expenditure was ¥350,400 million (FY 1999), ¥406,018 million (FY 2000), ¥390,892 million (FY 2001), ¥395,966 million (FY 2002) and ¥354,862 million (FY 2003). NTT Financial Reports (2000–2004).

14 The recent changes were noted in MIC Council R&D Report, p. 80. R&D expenditure used by the NTT holding company's R&D institutes, which are responsible for basic R&D, decreased from about ¥210 billion in FY 2001 to ¥160 billion in FY 2003. In 2003, the 'Comprehensive Commercialization Functions' scheme was established 'to facilitate smooth introduction of the results of R&D based on the better understanding market trends and commercial company needs'. See 'NTT R&D Revolution', *NTTis* (Spring 2004). As for NTT's emphasis on pure knowledge creation, see David T. Meth_, 'Living on the Edge: Basic Research and Knowledge Creation in Japanese Electrics Companies', in John R. McIntyre (ed.), *Japan's Technical Standards* (Westport, CT: Quorum Books, 1997, pp. 54–60).

15 The total R&D expenditure was ¥354,862 million, after subtracting the amount of internal transaction, ¥146.9 billion.

16 See, for example, NTT press release (24 October 2003) on Photonic Internet Lab. (NTT, NEC, Furukawa Electric and Mitsubishi Electric), which succeeds in interworking the next-generation photonic network control protocol of GMPLS and the advanced IP network control protocol of MPLS for the first time in the world.)

17 NTT DoCoMo press release (19 December 2003) on NTT DoCoMo's investment in FOMA handset development ('DoCoMo will provide a total of about 37 billion yen over two years to Fujitsu, Mitsubishi Electric, Motorola Japan, NEC, Panasonic Mobile Communications [Matsushita], and Sharp'. 'Under the terms of investment, DoCoMo will share ownership rights with the manufacturer for any new handset technology thereby smoothing the way for more efficient R&D in the future').

18 See, for example, 'Saraba NTT Siyou' (Goodbye NTT Specification), *Nikkei Communications* 297: 85–87, 89 (5 July 1999); Yuji Mizuno, *Hitachi* (Tokyo: Nihon Keizai Shimbun, 2004, pp. 234–237); PC Watch Impress (8 July 2003) [Fujitsu president's speech at news conference], http://pc.watch.impress.co.jp; NEC Financial Statement (June 2004) [Business risk of depending on the trade with NTT DoCoMo]. See also Kiyoshi Tsukamoto, *DoCoMo to au* (DoCoMo and au) (Tokyo: Kobunsha Publishing Co., Ltd., 2004, pp. 202–203) [interview with Panasonic director].

19 Under the law, the conditions have to be specified so as to ensure that: (1) the provision of telecommunications service shall not be extremely hindered by damage or failure of telecommunications [facilities] or telecommunications circuit facilities shall not be damaged, nor shall functions thereof be impaired [terminal equipment]; (2) the quality of telecommunications services shall

maintain an appropriate level [facilities]; (3) the secrecy of communications shall not be violated [facilities]; (4) the telecommunications facilities of users or other telecommunications carriers connected shall not be damaged or impaired, nor shall functions thereof be impaired, or not be suffered with any nuisance [facilities and terminal equipments]; (5) the demarcation of responsibilities between the telecommunications facilities of a telecommunications carrier and those of others [facilities] or between the telecommunications circuit facilities established by a telecommunications carrier and terminal facilities connected to them by a user [terminal equipments] shall be clearly stipulated (Articles 41 and 52).

20 The leading position of carriers in *de facto* standard-related activities in Japan still seems to be the case, which is pointed out in TTC Technical Study Advisory Group, Survey Report on Telecommunication-Related Forums' Activities (Version 9) (March 2003).

21 ITU-T subcommittee, MIC Telecommunications Council, report of 19 March 2003.

22 The number of NTT attendances at ITU meetings came to 188 from July 1998 to August 2000. The numbers of attendances of other major companies during the same period are as follows: Lucent 224; France Telecom 184; Siemens 153; Ericsson 152; BT 148; Nortel Networks 142; Deutsche Telekom 134; Alcatel 116; AT&T 77. As for Japanese companies, the numbes are: KDDI 69; NEC 47; Fujitsu 42; and Oki 32. ITU, 'Top Members' Participation from July 1998 to August 2000'. NTT Public Co., *Denki Tusin Zisyu Gizyutu Kaihatusi Hanso Denwa* hen (Carrier Telephony) (Tokyo: Telecommunications Association, 1971, ch. 7) and NTT Public Co., *Kokan hen* (Switch), p. 67 (1976) describe the situation in the early days.

23 See Hajime Yamada, *Gizyutu kyoso to Sekai hyozyun: Global Standard* (Tokyo: NTT Publishing Co., 1999, pp. 70–71).

24 MIC Council R&D Report, p. 119 recognized that the Japanese contribution to IEEE has been limited. Only 1.6 per cent of Requests for Comment, or standards set by IETF, and 3.7 per cent of Recommendations of W3C are estimated to be written by Japanese. MIC Institute for Information and Communication Policy, Report of Internet Study Group (September 2004).

25 The TTC and ARIB members are seeing standardization in Internet technology, mobile and security technologies and standard-setting organizations in areas such as IETF and IEEE as particularly important. MIC Council R&D Report, p. 164, appendix.

26 See Robert Galvin, *Nihon jin ni Manabi, Nihon ni Idomu* (Challenge Japan, Learn from the Japanese) (Tokyo: Nihon Keizai Shimbun, Inc., 2000, §4); Douglas J. Puffert, 'Technical Standards and Access to the Japanese Cellular Communications Equipment Market', in John R. McIntyre (ed.), *Japan's Technical Standards* (Westport, CT: Quorum Books, 1997, pp. 91–92).

27 See Tetsu Machida, *Kyodai Dokusen: NTT no Syukuzai* (Giant Monopoly: NTT's Remaining Wrong Doing) (Tokyo: Shinchosha Co., 2004, pp. 112–115). Softbank is considering entering the mobile market based on UMTS Time Division Duplex (UMTS-TDD) standards, which would become 3G mobile standards implemented in Japan. See Yukinobu Mizunoe, *Softbank Sansen de Kawaru Keitai Gyokai Seiryokuzu* (Foreseen Change the of Mobile Communications Landscape after Softbank's Entry) (Tokyo: Pal Pub., 2004, pp. 32–33).

28 MIC Council R&D Report, §2, and Ministry of Posts and Telecommunications Study Group Report on Globalization of Information Communication Technologies (5 June 1998) point out the need for Japanese companies to be committed in international standards against the background.

29 Japan Fair Trade Commission, 'Warning against NTT DoCoMo', 27 April 1999.

30 See JETRO, Survey on Actual Conditions, pp. 15–17.

31 See Tsukamoto, *DoCoMo to au*, §1. Another advantage is to utilize the broader global equipment market to cut the cost. See interview with Vodafone Japan CEO, *Nikkei Communications* 429: 64–65 (1 January 2005).

32 The applications by NTT West and NTT East are included, in view of their closeness. The number of published patent applications jointly applied by NTT, NEC, Fujitsu, Hitachi and Oki was 18 in 1994, 27 in 1995, 6 in 1996 and 23 in 1997. Since then, the number has been in the half-dozen range.

33 As for the status in the mid-1980s in the cable industry, see Fransman, *Japan's C&C Industry* p. 239. The joint patent ownership practice was a compromise after hard negotiation in the early days. Later, it was noted as a success factor in the epoch-making crossbar switch developed by NTT Public Co. and its family. NTT Public Co., *Danki Tusin Zisyu Gizyatu Kaihatusi*, p. 370.

34 MIC Council R&D Report, p. 128, points out that service operators are generally more willing than are manufacturers to refrain from claiming their share from the patent that was implemented in the standards.

35 For details, see ITU-T and ITU-R Patent Policies and their declaration forms, available at http://www.itu.int/ITU-T/dbase/patent/index.html and http://www.itu.int/ITU-R/study-groups/patents/index.html. The data were taken from ITU-T and ITU-R patent databases as of May 2004 (ITU-T) and March 2005 (ITU-R). The number of licensing declarations submitted to ITU-R regarding broadcasting technologies is not counted in the figures. 'General' in figures notes the number of submissions of a general licensing declaration, which declares the willingness to license in the event that part(s) or all of any proposals contained in contributions submitted by the organization are included in ITU-T Recommendation(s) and the included part(s) contain items that have been patented or for which patent applications have been filed and whose use would be required to implement ITU-T Recommendation(s). Note that the IPRs policy practice and patent statements are continuously changing. The data and description reflect the status at the date noted. ITU-T revision history can be seen in 'README Information about the Guidelines for ITU-T Patent Policy and about the Patent Statement Declaration Form' (http://www.itu.int/ITU-T/dbase/patent/readme-guide.html) (March 2005).

36 Note, however, that it is difficult to verify whether NTT's royalty rate is indeed reasonable, as information on the royalty rate is kept secret between the licensors and licensees, and it is also difficult to define the reasonability. Setting licensing policy according to the region is a common practice, and NTT's behaviour can be different outside Japan. Jeffrey L. Funk, *Global Competition between and within Standards: The Case of Mobile Phones* (London: Palgrave, 2002, p. 56), indeed notes that the high licensing fee levied by NTT on mobile equipment manufacturers inhibited promotion of their second-generation equipment abroad.

37 As well as this, NTT has not been a patent licensee. NTT is basically a procurer that sets up the network and uses it for its services with the equipment procured from the manufacturers. According to settled principle under the patent law in Japan, when the manufacturers sold the patented item to NTT, the patent had been understood to be exhausted, and the patent owner could not claim any right against the buyer, NTT. Where there is any risk of infringement, NTT has immunity from the manufacturers, which has had to take care of any risk of infringement. An exception to this principle might be brought by the MPEG-LA, a successful patent pool based in the United States, in which a licensing programme planned to charge users of the patented equipment

royalty calculated based on their usage ('use fee'). Japanese service providers have not been familiar with the idea and are strongly opposing it. There have been no published patent infringement cases brought by NTT, which might show NTT's generosity to some extent, though a low case profile often just indicates a patent owner's preference for settlement.

38 See, for example, Yamada, *Gizyuto kyoso to Sekai hyozyun*, §6.4.

39 Other mitigating factors are observed in Japanese trade and corporate governance practices. Under Japanese corporate governance, companies have been allowed to invest and conduct R&D from long-term perspectives, which could be a reason for the low profile of patent enforcement. Other stabilizing factors will include a long relationship between companies in several markets. In a Japanese company, the division that sold its product to a patent infringer, or potential licensee, of the company tended to oppose the enforcement to them, through a fear of losing their customer because of the enforcement. This mitigating factor in patent enforcement is considered to have been working well in Japan, as Japanese companies in many fields have often sold some of their products to others.

40 Federal Trade Commission (USA), *To Promote Innovation: The Proper Balance of Competition and Patent Law and Policy*, ch. 3, p. 30 (October 2003).

41 Now the government is keenly trying to promote technology-based start-ups. See, for example, Venture Support Centre, National Institute for Information and Communications Technology (NICT), http://www.nict.go.jp/overview/section/jyohosinko_e.html (March 2005). Other potentially 'onerous' patentees would be the academics. The dedicated patent portfolio setting-up service is growing in the United States and the United Kingdom, which are promoting their services in Japan. The changes that might be caused by this trend are yet to be seen.

42 In this regard, 'Reference Guideline for Intellectual Property Information Disclosure' is published by the Ministry of Economy, Trade and Industry on 27 January 2004, which is expected to promote IPRs-related information disclosure to investors.

43 See Intellectual Property Policy Headquarters, The Cabinet, Intellectual Property Promotion Plan 2004 (27 May 2004).

44 The early case that made Japanese companies realize the seriousness of the problem was Motorola's patent claim for a G3-type facsimile modem after the standard had been set at the International Telegraph and Telephone Consultative Committee (CCITT, predecessor of ITU-T) and implemented in Japan. See Kazunori Ishiguro, *Densi Shakai no Ho to Keizai* (The Law versus Economics in Electronic Society) (Tokyo: Iwanami Shoten, 2003).

45 MIC Council R&D Report, p.128.

46 The 'licensing statement' is submitted to ARIB under its Patent Policy, which is explained in detail on p. 168.

47 See also Japan Patent Office, *Patent Map: Mobile Communication System* [in Japanese], (Tokyo: Japan Institute of Invention and Innovation, 2000, p. 33) [Trend of patent applications during 1986–1996].

48 See note 17.

49 See Carl Shapiro, 'Navigating the Patent Thicket: Cross Licenses, Patent Pools and Standard Setting', in Adam B. Jeffe, Josh Lerner and Scott Stern (eds) *Innovation Policy and the Economy*, vol. 1 (Cambridge, MA: MIT Press, 2001, pp. 119–150).

50 See Intellectual Property Promotion Plan (note 43), §3.2 (3). Although the issue is discussed in a wider context, not limited in the telecommunications sector, the information communication and telecommunications sectors are recognized as the most relevant sector, as well as bio-industry. Particularly for the status in telecommunication sector, see MIC Council R&D Report.

51 The notable exceptions include W3C, which adopted the policy that only royalty-free technology could be implemented into its standards. It is said, however, that we cannot expect the same approach from other SSOs of like size. W3C indeed maintains a peculiar culture whereby individuals, academics and/or independent small technology-oriented companies, who often support the philosophy that the Internet should never be under anyone's control, retain a powerful voice. It is even said that the fact that even W3C had to go through a hard discussion to reach the conclusion would suggest the difficulty for a large international standard-setting forum to have a policy of this sort. More focused forums, such as the Bluetooth standard-developing forum, and individual companies that are controlling *de facto* standards, occasionally implement a royalty-free licensing policy. But these are on a rather *ad hoc* basis and are possible only when those involved believe the approach is inevitable to promote their work.

52 The Telecommunications Technology Committee (TTC), 'Policy for the Handling of Industrial Property Rights', 18 May 1989, amended on 24 October 2001, 28 March 2002 and 3 March 2003; TTC IPR Committee, 'Operation Procedures for the Handling of Industrial Property Rights', 27 May 2002, amended on 10 December 2002 and 17 April 2003. An English version is available at http://www.ttc.or.jp/e/intro/rules/ru6/index.html (accessed March 2005). The submitted statements can be seen on TTC's web page, http://www.ttc.or.jp/j/ipr/index.html [in Japanese]. The Association for Radio Industry Businesses (ARIB), *Hyozyun kikaku ni kakaru Kogyo syoyuu ken no Toriatukai ni kansuru Kihon hosing* (Policy for the Handling of Industrial Property Rights), 5 September 1995 and its Operational Policy, 5 September 1995, amended on 29 March 2000, http://www.arib.or.jp/tyosakenkyu/sakutei/sakutei04.html [in Japanese].

53 See Toru Yamauchi (Ministry of Economy Trade and Industry) 'A Perspective on Standardization and Intellectual Property', in Sherrie Bolin (ed.) *The Standards Edge* (Menlo Park, CA: Bolin Communications, 2002, p. 173).

54 Presentation of Prof. Sadao Nagaoka at Competition Policy Research Centre Seminar, Japan Fair Trade Commission (18 February 2005).

55 See Intellectual Property Promotion Plan (note 43), pp. 68–69.

56 See, for example, Isamu Yoshimatu and Hajime Yamada, 'ITU-T IPR Ad-hoc Group Meeting' [in Japanese], *ITU Journal* 34 (10) pp. 24, 26 (October 2004).

57 Japan Patent Attorneys Association Central Research Institute of Intellectual Property, *Gizyutu Hyuzyun to Tokkyo ken ni tuite* (Technical Standards and Patent), p. 52 (January 2005) proposes involvement of the Japan Intellectual Property Arbitration Centre, which was established by Japan Federation of Bar Association and Japan Patent Attorneys Association. A comprehensive study in the earlier days suggested considering the use of WIPO Arbitration and Mediation Centre. Institute of Intellectual Property, Study Report on IPR issues concerning Technology Standards [in Japanese] (Tokyo: March 1995).

58 To limit the standard to be under thorough IPR analysis will help cut costs; it would not always be easy to select the standard by any criteria. It was also rightly pointed out that companies do not always conduct a thorough, time-consuming patent search and analysis; they conduct a search only to the extent necessary to eliminate the risk of being ordered to stop their business and pay an unjustifiable amount of damages. This approach helps to cut the cost of patent searches, but would be difficult to implement by the SSOs, which do not (or should not) know the future business plan of the licensee and, accordingly, cannot assess the risk of infringement and the importance of getting a licence. As for the patent search, the secrecy of patent applications under the US

patent system was an obstacle, as it put employees involved in SSOs in a difficult position: they are not allowed to disclose the secretly kept information about their company's patent applications, but at the same time are asked by SSOs to disclose it. The difficulty has been eased by the US introduction of the mandatory application disclosure system in 2001. A remaining serious impediment to patent searches is the US treble damage charged against wilful infringement, which puts the company that informs on others' patents at risk of treble damage. Although telecom service providers, which are relatively free from the risk of infringement, are in a good position to inform on the others' patents, manufacturers, who actually conduct more thorough patent searches, usually, put themselves at serious risk by informing on others' patents. Another practical concern is involved in the incentive to participate in SSOs. It is commonly thought that having a strong IPRs policy is damaging for participating companies. The *de jure* telecommunications SSOs have been relatively free from this worry because of the recognized need for the companies to be involved in telecommunications standard-setting because of telecommunications' wide effect on society and because interchangeability, interoperability and interconnectivity are critical to any telecommunications operators and manufacturers that are running material-sized business. These conditions still hold; however, recourse to the *de facto* standard is increasingly seen. More importantly, a strong IPRs policy might keep technology-oriented companies, that are not involved in services and manufacturing, out of the SSO for fear that their core asset might lose its effectiveness through involvement.

59 See Intellectual Property Promotion Plan (note 43), p. 69; Council for Science and Technology Policy, The Cabinet, Recommendation on Intellectual Property Strategy [in Japanese], p. 16 (19 June 15). See also Japan Patent Attorneys Association Central Research Institute of Intellectual Property, *Gizyutu Hyuzyun to Tokkyo ken ni tuite* (Technical Standards and Patent) (January 2005); Japan Federation of Economic Organizations, *Senryaku teki na Kokusai Hyozyunka no Suisin ni kansuru Teigen* (Position Paper on Promotion of Strategic Global Standardization) IV, p. 30 (January 2004); Yamauchi, 'A Perspective on Standardization'.

60 For details, see Larry M. Goldstein and Brian N. Kearsey, *Technology Patent Licensing* (Boston, MA: Aspatore, 2004).

61 For details, see Masako Wakui, 'Standardisation and Patent Pools in Japan', in Ruth Taplin (ed.) *Valuing Intellectual Property in Japan, Britain and the United States* (London: RoutledgeCurzon, 2004, pp. 81–109).

62 See, for example, Hidehisa Tanaka, Toru Takaya and Ayato Susaki, 'Standarization and IPR: Global Situation, Japanese Challenges', in Bolin, *Standards Edge*, p. 179.

63 Interview with Japanese companies officials (October 2004–December 2004).

64 Patent Law Article 93 [Arbitration decision on the granting of a non-exclusive licence in the public interest] specifies that: (1) where the working of a patented invention is particularly necessary in the public interest, a person who intends to work the invention may request the patentee or the exclusive licensee to hold consultations on the grant of a non-exclusive license; (2) if no agreement is reached or no consultation is possible under the preceding subsection, a person who intends to work the patented invention may ask the Minister of Economy, Trade and Industry for an arbitration decision. In some cases, article 83 [Arbitration decision on the granting of a non-exclusive licence in the case of non-working] and Article 92 [Arbitration decision on the granting of a non-exclusive licence on one's own patented invention] may also be relevant.

65 Report of Working Group on Patent Strategy Plan, Patent System Subcommittee, Intellectual Property Committee, Industrial Structure Council, Ministry of

Economy, Trade and Industry (METI), 'Issues Concerning Smooth Use of Patented Inventions' [in Japanese] (17 November 2004). The study was organized at the Technology Research Division, General Affairs Department, Japan Patent Office.

66 See Japan Intellectual Property Association, Position Paper on Compulsory Licence, 21 September 2004 [in Japanese], reprinted in Report of Working Group, ibid.

67 See Intellectual Property Promotion Plan (note 43), p. 68.

68 Report of Study Group on Technical Standards and Competition Policy, Japan Fair Trade Commission (25 July 2001).

7 Policy network for network policy in Japan

Kenji Suzuki

Introduction

Telecommunications policy requires great technological expertise, but this does not mean that the policy stands purely on technological grounds. On the contrary, it seems to be quite vulnerable to political intervention. There are several reasons for this. First and foremost, telecommunications makes a great contribution to the national economy and society, such as economic growth, social welfare, defence, and so forth. Telecommunications is also a big business in its own right. The operators are far from negligible as employers, buyers and taxpayers. The large number of employees also implies that telecommunications policy should affect many votes for politicians.

The complex and ever-changing nature of telecommunications technology allows room for political manoeuvre. For instance, how to promote competition in the telecommunications market is one of the questions that is often very significant for policy-making, but there is no consensus on the answer to it. The lack of consensus often induces intervention from other interests. Technological innovation causes political conflict also by breaking the traditional order of regulation. The development of international communication systems has reduced the significance of national regulation, while increasing the potential for trade conflict and cross-border mergers and acquisitions. The convergence of telecommunications and electronic products as a result of digitization may cause regulatory disturbance where those industries are put under different jurisdictions.

Consequently, telecommunications policy is one of the most popular subjects in the literature of political science, not least in Japan.[1] This chapter contributes to the literature by providing an analysis from a perspective that has not yet been applied to the Japanese telecommunications policy: the perspective of policy network. To introduce the concept of network is useful in understanding complex situations of policy-making process in a systematic manner. For chronological observation of the network, moreover, the chapter focuses on a specific policy issue that has long been discussed – the organizational reform of Nippon Telegraph and Telephone (NTT) – instead of encompassing other issues.

The chapter is organized as follows. The next section presents the theoretical background and establishes the analytical framework for the present study. The third section investigates the network situation in the early 1980s, when the public telecommunications operator, NTT, was privatized and its organizational reform started to be discussed. This is followed by a fourth section that illuminates the subsequent transformation of the policy network. The final section summarizes and concludes the discussion.

Analysis of policy network

The policy network perspective became popular in political studies in the late 1980s, especially in the European literature. There are many variants of the perspective, but it is often applied to stress the importance of the interaction among various state and social stakeholders in the decision-making of particular public policies. Initially, it reflected a criticism of the traditionally mainstream models in the political science: pluralism and corporatism. As Rhodes and Marsh pointed out, 'Neither model provided a very realistic picture of the relationships between government and a realistic picture of the relationships between government and interest groups, largely because they purported to offer a general model of these relationships'.[2] Furthermore, many scholars applying the policy network perspective share the idea that the relationship between relevant stakeholders in the policy-making process is quite variable across different policy fields. No doubt macro-level politics should be considered in explaining any policy outcome,[3] but it is often difficult to attribute particular outcomes directly to macro-level politics. Even if pluralist/corporatist models are adapted to lower levels such as 'meso-corporatism',[4] their capacity to describe political structure is considerably limited because they only provide two particular sets of criteria, which are not even clear alternatives. It is possible to discuss whether or not their subjects fit in with the models of pluralism/corporatism, but it seems difficult to go further than that. In this respect, it is noteworthy that Muramatsu and Toyonaga provided excellent analyses of Japanese telecommunications policy, applying pluralism and corporatism respectively,[5] but they had to introduce such new concepts as 'patterned' pluralism and a 'bipartisan' approach to describe the deviation from the basic models. By contrast, policy network is 'a generic label embracing the different types of relationship state/interest group that exist in the process of determining any individual policy output'.[6] In other words, the policy network perspective allows researchers to understand the characteristics of the policy-making process without inventing new labels.

However, there is no consensus as to which network properties should be considered to characterize the policy-making process. While previous political studies have tended to provide many dimensions for typology,[7] this study takes a rather restrictive position to avoid complexity and focuses on two concepts: clique and centrality. The terms are taken from the literature of sociological/mathematical network studies,[8] but they are defined more subjectively and qualitatively so as to fit in with the analysis of the policy-making process.

In this study, a clique is defined as a group of stakeholders which are tied relatively closely within the network. A clique may contain the stakeholders with the same idea of policy output, and in this case it is identical to what Sabatier calls an 'advocacy coalition'.[9] However, a clique may also be formed regardless of policy ideas. A party may make a coalition with others just because it owes some resources to them. Centrality denotes the extent to which each stakeholder lies near to the centre of the network. To put it another way, the most central stakeholder has the largest influence on the decision of the policy output.

In this study, we focus on the chronological change of policy network. The hypothesis underlying this approach is that the Japanese telecommunications policy network is not very stable, reflecting the absence of an 'iron triangle'. 'Iron triangle' denotes a closed policy network where bureaucrats, legislators and social interest groups establish a close relationship. The three groups share compatible goals and support each other. Members of each group are usually strongly united, hence the policy-making process is often viewed as the interaction between the representatives of those groups. The iron triangle concept is often mentioned to describe the policy-making process in Japan, but Japanese telecommunications policy is not a case in point. This has been pointed out by previous studies,[10] but it has not been discussed in relation to the dynamics of policy network.

Our observation encompasses the period from 1981. To elucidate the chronological change, this study focuses on the policy discussion at four different stages. The first stage is the organizational reform initiated by Ad Hoc Commission on Administrative Reform (*Rincho*) from 1981 to 1984, leading to the privatization of NTT. The second stage is the discussion on the break-up of NTT from 1988 to 1990, which did not reach any conclusion. The third stage is the resurgence of that discussion, which resulted in the formal break-up of NTT, but with the ownership preserved by the sole holding company. The fourth stage is a series of discussions on telecommunications policy from 1999 to 2003. The basic structure of the telecommunications policy network at the first stage is identified in the next section. The fourth section explores the following three stages, illuminating how the network structure has been transformed over time.

Telecommunications policy network at the initial stage: the policy-making process towards the privatization of NTT

Relevant stakeholders and their interests

The privatization of NTT was special in the sense that it was not discussed alone, but as a part of a wider framework, *Rincho*. *Rincho* was established in 1981, with the aim of large-scale administrative reform. Among other changes, the organizational reform of the three largest public corporations, including NTT, drew much attention from society.

The performance of NTT was not as bad as that of the other public corporations, notably Japan National Railways. Nevertheless, there were several signs that NTT would soon get into trouble. Excess labour was conspicuous, as the number of the employees did not fall even after the accomplishment of universal service in 1977 and the full automatization of the operation system in 1978. The managers could not handle the problem, not only because they were not very business-minded, but also because of lack of flexibility in the management system regulated as a public corporation. For instance, managers could not fix their budget. The salary system was fixed at the same level as for other public corporations. There was no link between salary and achievement, and hence the moral hazard of the employees became conspicuous.

The solution of the problem provided by *Rincho* was the simultaneous privatization and break-up of NTT. According to a former staff member, the ultimate objective of *Rincho* was to secure market competition in the telecommunications sector.[11] Therefore, *Rincho*'s final report in 1982 recommended not only the privatization of NTT, but also its break-up, so that NTT would lose dominant power and new operators could enter the market. At least apparently, the interest of *Rincho* was based on pure economic considerations and did not reflect other interests. The recommendation was quite consistent with the position of the economics professor Hiroshi Kato, who chaired the *Rincho* subcommittee in charge of NTT. As discussed later, furthermore, *Rincho* was quite independent from bureaucrats and politicians.

The *Rincho* recommendation elicited a reaction from other stakeholders. NTT itself submitted three alternative plans (gradual reform under the existing framework, reform into a semi-public organization, and privatization) to *Rincho*. They were different in terms of ownership, but none of them favoured organizational break-up. NTT presented three alternatives not because it was so flexible that it felt able to give options to the government, but because it could not obtain a consensus. In fact, there was a division of opinion between a conservative group led by the vice-president, Yasusada Kitahara, and a progressive group supporting the president, Hisashi Shinto. The conservative group opposed privatization because if it went ahead they would not retain the prestigious status of working of a public authority, and also because they were concerned that the pressure of market competition would set back basic research and other long-term investment. Whereas Kitahara was promoted from within, Shinto was appointed from outside. Unlike his predecessors, he did not even have a career in the bureaucracy. Instead, he was a famous businessman and well regarded as a 'cost-cutter' for his previous work as the president of Ishikawajima Harima Heavy Industries, an entirely private company. It is therefore not surprising that he emphasized the positive aspects of privatization, such as the autonomy of management and the flexibility of wage-setting, rather than the negative aspects. Shinto held a positive view on the break-up of NTT, at least in the early stages. According to Suzuki,[123] Shinto informally agreed to *Rincho*'s idea that privatization and break-up should not be conducted separately, and even considered breaking NTT up into 47 separate entities, each corresponding to one prefecture.

By contrast, the trade union of NTT, the Japan Telecommunications Workers' Union (Zendentsu), was a formidable opponent of privatization. It is true that the union leaders preferred the liberalization of wage-setting, not least because they feared that their wages would stop growing, because of the financial problems of the other public corporations. Nonetheless, they believed that the problem could be overcome without any change of ownership. Hence, they proposed yet another alternative, joint ownership by the government and the subscribers. However, Zendentsu's opposition to privatization was not so strong as its opposition to break-up. If NTT were broken up and the working conditions were differentiated from one company to another, the unity of Zendentsu would be at risk. This was not acceptable, since Zendentsu's power was based on its large size.

Another important stakeholder was the Ministry of Posts and Telecommunications (MPT). Although MPT was formally in a position to supervise telecommunications policy and NTT, its role until the 1970s was no more than liaison between NTT and the Diet. Historically, NTT and MPT were the colleagues of the same ministry, the Ministry of Communications. When they were divided in 1949, almost all the workers in charge of telecommunications technology went to the Ministry of Telecommunications, which later became NTT. From NTT's point of view, therefore, MPT had no knowledge of, and should not intervene in, telecommunications policy. Indeed, MPT had only two officials in charge of the supervision of NTT, one of whom was successively seconded from NTT itself.

However, as it was widely recognized that telecommunications was a key factor for technological innovation and economic growth, MPT was interested in taking the initiative in telecommunications policy. It established a new bureau for communications policy in 1980. Whereas MPT had hitherto been regarded as an 'operational agency', it was now intended to turn it into a 'policy agency', which would be more respected in the bureaucratic community.

Against this background, MPT was opposed to the organizational reform of NTT. To respond to the *Rincho* recommendation, for instance, the chief of the Communication Policy Bureau stated that 'I recognize the need to maintain telecommunications as a public sector and to extend it qualitatively and quantitatively to prepare for the coming information society'.[13] The underlying concern was that privatization would give too much freedom to NTT. As a private company, it would become selective in its operations, abandoning less populated areas owing to their low profitability. Furthermore, it would stop capital investment, which is beneficial in the long run but is not profitable in the short run. All profits would be used up to pay the salaries of inefficient workers and managers... In short, MPT feared the loss of public control over the giant corporation. While MPT was opposed to the *Rincho* recommendation, its logic was quite different from that of NTT.

According to Vogel, however, the attitude of MPT was not monolithic.[14] There had been the idea that privatization would be rather beneficial for the ministry. If NTT were to be privatized, it would no longer act as a public authority. Therefore, MPT would be able to play an exclusive role in telecommunications policy. Moreover, the ministry would be able to extend its authority through the entry of new operators in the telecommunications market. In fact, even before the launch of *Rincho*, MPT discussed the organizational reform of NTT in the Telecommunications Policy Discussion Group, a semi-formal advisory body established under the Communication Policy Bureau. Because of this, MPT was rather suspected of wishing to promote the organizational reform of NTT.[15] On balance, it may be fair to say that MPT was opposed to the reform but not in a determined manner, being prepared to change position for the sake of its main interest: that of maximizing its own authority.

The Ministry of International Trade and Industry (MITI) was cautious about the expansion of MPT's authority. The ministry envisaged enlarging its authority to cover telecommunications as an extension from the electronics industry. When MPT proposed to establish the Communication Policy Bureau, therefore, MITI demonstrated strong opposition. The two ministries also had battles on such issues as value-added networks and test sites for multimedia networks. However, MITI apparently did not join in the debate over the organizational reform of NTT, presumably because it was not clear how it affected the regulatory scope of the MPT.

Rather than MITI, it was the Ministry of Finance (MOF) that had a strong interest in the NTT issue. While it was not concerned about the management of NTT or the telecommunications market, it was anxious to privatize NTT for financial reasons. For MOF, the privatization of NTT meant an extraordinary windfall for the government. After a decade of low growth, the financial deficit of the Japanese government had become very serious. It should also be noted that the government failed to introduce value added tax in 1979, and there was no other option for increasing budget revenue at that time. From the viewpoint of MOF, however, the *Rincho* recommendation was too much, in the sense that it recommended the break-up of NTT. NTT should be kept in one unit to maximize its share value.

For politicians, the governing Liberal Democratic Party (LDP) was internally divided into several groups on the NTT issue. To understand this, it is necessary to be aware of a special concept that was invented to explain Japanese politics: '*zoku*' ('tribe' in English). *Zoku* represents a group of politicians acting as a patron of bureaucrats and other interests in a particular policy area. The members were often very knowledgeable about the policy area, as they successively held such positions as the minister, the parliamentary vice-minister, a member of the Diet standing committee, and the member of the party research council for policy affairs. Through those careers, they also developed a close out-of-Diet connection with the bureaucrats and other interest groups. Naturally, they often played a crucial role where an 'iron triangle' was formed between politicians, bureaucrats and social interests. According to the data compiled by Inoguchi and Iwai, there were eleven *zoku* groups, including 282 members in the mid-1980s, which accounted for over 60 per cent of the party Diet members.[16]

Reflecting the interests of MPT and MOF in the NTT issue, the patrons of those ministries, respectively called *Teishin-zoku* and *Okura-zoku*, tried to lead party opinion in favour of their ministries. Yet there was another important group in the discussion: the supporters of *Rincho*. They supported *Rincho* either because they were truly reform-minded, or because they wanted to support Yasuhiro Nakasone – a strong supporter of *Rincho*. He was the Minister for Administrative Management and took the initiative in the establishment of *Rincho*. He became Prime Minister after that, raising administrative reform as the key mandate of his cabinet. The interest of LDP was structured by the interaction of those three groups.

For the opposition in the Diet, the largest opposition party, the Socialist Party, was constantly opposed to the *Rincho* recommendation. Presumably this was not only because the party was sceptical about liberalization and deregulation, but also because it had calculated that the defeat of the *Rincho* proposal would cause serious damage to the credibility of Nakasone and his party.

Cliques and centrality of the Japanese telecommunications policy network

Policy stakeholders often ally themselves with each other because they share the same policy position, and/or because they are indebted to each other. Consequently, a number of cliques emerge in the policy network. In the case of the Japanese telecommunications policy in the early 1980s, there were four such cliques present in the network.

Rincho was at the core of the clique that pushed for both the privatization and the break-up of NTT. Although it was only a temporary advisory body, it enjoyed strong support from the public, who were dissatisfied with NTT's ineffective management and the unduly good conditions for its workers. This public support was a very attractive resource for the top LDP leaders, especially the Prime Minister, Nakasone, who identified himself as the leading administrative reformer. *Rincho* was put under the direct responsibility of the Prime Minister, and it was not possible for other ministries or their patrons to intervene in its decision-making.

Rincho was also backed by many businessmen. This was because of its chairman, Toshio Doko, who was one of the most prominent businessmen in the postwar period. He was also a former chairman of the peak business association, the Japan Federation of Economic Organizations (Keidanren). The appointment of Doko and some other top businessmen to the *Rincho* board was very likely an important factor in drawing support from a large part of the business community. Many companies supported *Rincho* because they also saw the liberalization of the telecommunications market as a great opportunity for their own businesses.

The chairmanship of Doko is also important when the position of Shinto, the president of NTT, is considered. Doko and Shinto had built a strong personal link as they worked together at Ishikawajima Harima Heavy Industries. Indeed, Doko was one of the people who recommended the government to appoint Shinto to be the president of NTT.[17] Shinto claimed that Doko did not refer to the organizational reform of NTT when he was appointed,[18] but it was not surprising that Shinto became sympathetic to *Rincho*. The connection with Doko also seemed to give him stronger bargaining power against his opponents from inside. According to several NTT officials interviewed by Vogel, 'they felt that Shinto had the blessing of both Keidanren and the *Rincho*, and thus to oppose him would be futile'.[19]

Besides the *Rincho* clique, another clique was formed around the conservative group of NTT officials. The membership of this clique included a part of the business community. While many companies supported *Rincho*, a number of companies strongly opposed it. Most of them were the suppliers of telecommunications equipment to NTT. It is perhaps understandable that they opposed the reform, since it would affect the market they had monopolized. However, the liberalization of the telecommunications market would also give new opportunities and it was not necessarily bad for them. An additional, and perhaps more powerful, explanation for their opposition is that they had a special connection with NTT in such a way that they identified themselves as the 'NTT family'. The partnership between NTT and those companies was more than just a buyer–supplier relationship. They depended on the procurement of NTT, while NTT depended on their technology. There were various personal links ranging from top management to daily transactions. The most symbolic was that they constantly offered good job posts to retired NTT staff – so-called *amakudari* ('descent from heaven'). Market liberalization might well destroy all such historical efforts of connection-building. In other words, they benefited from the exclusive link with NTT under the existing system, and preferred to keep it rather than to seek new opportunities.

The conservative clique was also joined by the NTT trade union, Zendentsu. The membership of Zendentsu significantly increased the political power of this clique. Backed up by very high participation rate (nearly 100 per cent) of about 300,000 workers, Zendentsu was one of the most powerful trade unions in Japan. Zendentsu had its closest relationship with the Socialist Party, which backed this clique as well.

MPT also opposed the *Rincho* recommendation, but it is better not to see the ministry as a participant in this clique. MPT had a different logic for its opposition and was not so determined as other opponents. On top of that, there was strong tension between MPT and NTT for the initiative in telecommunications policy. They were unlikely to form an alliance even though they were both opposed to the reform.

Yet MPT did not stand entirely alone. It enjoyed strong political power backed up by *Teishin-zoku. Teishin-zoku* was one of the most powerful groups in the LDP at that time, for two reasons. First, *Teishin-zoku* had a strong leader, Kakuei Tanaka. He was a former Prime Minister and led the largest faction of LDP, which accounted for one-third of the LDP Diet members. He was considered the most powerful figure in the LDP even though he had no formal position in the party, as a consequence of a scandal related to his private dealings. The majority of *Teishin-zoku* were the members of the Tanaka faction. Indeed, it seems that Tanaka deliberately appointed the members of his faction to the important posts of post and telecommunications policy.

The second reason for the power of *Teishin-zoku* was that the group held the important function of collecting votes for the party – encouraging MPT to mobilize a large number of votes through the franchised postmasters. They were strongly united through the national organization, which often acted as a powerful lobby group. Furthermore, many of the postmasters (especially those in rural areas) were local dignitaries and allegedly had a strong influence on the voting pattern of their locals.

The last clique to be mentioned was formed by MOF and its political patron, *Okura-zoku*. They did not ally themselves with others because they were interested in a different goal: privatization without break-up.

The four cliques formed gradually, and policy discussion reached a critical point in summer 1983, when the LDP discussed its reform proposal to respond to the *Rincho* recommendation. In July, for instance, Shin Kanemaru, the number two of *Teishin-zoku*, proposed semi-public ownership of NTT. The final proposal of the LDP, published in September, concluded that NTT should be privatized but not broken up immediately, and that its break-up needed to be discussed ten years after privatization.

This result may be interpreted in two ways. First, the LDP proposal was the product of a compromise between the progressive clique and the conservative cliques. Second, the MOF/*Okura-zoku* clique had a dominant power over others, and led the LDP in the direction that was most convenient for them. Those two interpretations are not incompatible, and the truth was perhaps a combination of the two. The first three cliques may have been prepared to compromise with one another, when the last clique pushed the process forward. As Vogel has stated, 'behind the scenes the MOF was the real force driving the policy'.[20]

Of a number of individuals who brokered different cliques, the most prominent was Kakuei Tanaka. He was a *zoku* leader and a faction leader, but he was the leading figure of the LDP and needed to consider the interest of the party as a whole. Therefore, he had understood the necessity for reform even while apparently supporting his MPT/*Teishin-zoku* clique. At later stages, he persuaded his clique to make compromises, instead of using his power to hold down others. He also contacted the people at *Rincho*, including Doko and Kato, to signal his intention.[21] The leader of Zendentsu, Akira Yamagishi, also acted as an important policy broker. He was allegedly prepared to make compromises from an early stage in order to prevent a more radical reform. It was Yamagishi who persuaded Shinto to compromise and drop the idea of an NTT break-up.[22] Fujii also reported that Yamagishi had one-to-one contact with Nakasone and persuaded him to give up the idea of break-up.[23] On the other hand, he persuaded his trade union members to accept privatization.

Subsequent transformation in the telecommunications policy network

The second stage, 1988–1990

As a result of the policy discussions, NTT was privatized in 1985. The corporation was not broken up at that time, but it was decided that the government should make a decision on organizational reform within five years – that is, by the end of March 1990. The former participants of the telecommunications policy network started to discuss the issue in 1988, when MPT consulted the Telecommunications Council. However, several major participants were missing by that time. *Rincho* had already been dissolved after its final proposal in 1983. Doko passed away in 1988. Nakasone left office in 1987. Shinto successfully became the first president when NTT was privatized in 1985, but he left NTT in 1988 following a private dealings scandal. As a result, the progressive clique that advocated the break-up of NTT was not part of the policy network this time.

There were several changes in other cliques. While NTT had previously been divided into two groups, now it was united in the conservative position. This is partly because Shinto's successor, Yamaguchi, was not so reform-minded as Shinto, and partly because a large majority of officials did not think that organizational break-up would be necessary. The downfall of Shinto also affected the relationship between NTT and the NTT family companies. Shinto tried to bring in new business partners to depart from the traditional closed relations with a limited number of family companies. However, due to his private dealings with one of those new partners, NTT became cautious in making new partnerships. The traditional relationship was retained, and the family companies were still interested in giving support to NTT.

Zendentsu also stayed in this clique. It did not oppose the self-reform plan of NTT, which proposed a 15 per cent reduction in the number of employees (from 270,000 to 230,000) over five years. Obviously, the plan was published in order to avoid public criticism of the performance of NTT, which would lead to organizational reform. As a trade union, Zendentsu might well be concerned about workforce reduction, but the union leaders seemed to prioritize the prevention of break-up.

On the other hand, MPT had become a strong promoter of the break-up of NTT by 1990, whereas it took a conservative position in the first reform discussion. This was not because the ministry had changed its interest. Its goal remained the extension of its authority to turn it from an operational agency into a policy agency. Since NTT was already privatized, it would be easier for MPT to control smaller pieces than a giant corporation. The break-up of NTT was also a demand of the new common carriers (NCCs). Unlike NTT, NCCs were very compliant to MPT. They offered job posts to retired MPT officials in order to keep good contact with the ministry. Together with *Teishin-zoku*, MPT and NCCs formed an 'iron triangle'. Against this background, MPT was willing to support the interest of NCCs. Reflecting the interest of MPT as such, the Telecommunications Council proposed at the beginning of March 1990 that NTT should divide its business between regional operations and long-distance operations to ease the difficulty of NCCs in competing with NTT in the long-distance telecommunications market. The Telecommunications Council is formally an institution composed of academics and other specialists to advise the post and telecommunications minister, but it was substantially controlled by the ministry officials and used as an authorizer of the argument of the officials in the ministry. The Telecommunications Council was cautious about the division within regional operation, and this was considered a strategy by MPT to focus on the most urgent matter to achieve the reform quickly.

An important feature of the telecommunications policy network at the second stage was the exposure of the conflict between the two cliques mentioned earlier: the NTT clique (NTT, the NTT 'family' and Zendentsu) and the MPT clique (MPT, *Teishin-zoku* and NCCs). The traditional conflict of interest between NTT and MPT over the initiative of telecommunications policy was now translated into the conflict over the organizational reform of NTT.

The LDP seemed to be more sympathetic to MPT just after MPT published its proposal, presumably because of the lobbying of *Teishin-zoku* within the party. For instance, the chairman of the party research council for telecommunications policy, Tsutomu Hata, stated that the division between regional operation and long-distance operation 'was an appropriate method and there was no significant opposition within the party'.[24] The MPT clique stood in a better position, since it referred to Shinto's scandal as a consequence of NTT's market dominance.

Nevertheless, the NTT clique obtained strong support from MOF. The ministry reduced its share of NTT by 34.6 per cent between 1986 and 1988, but it then stopped selling owing to the downturn of the stock market. It therefore watched for the next chance to sell. From the viewpoint of MOF, the division of NTT was only harmful, as it seemed to reduce the asset value of the corporation. In fact, NTT's share price fell after MPT published its proposal. The finance minister expressed his strong opposition in the cabinet, and the ministry declared that, as the largest shareholder, it would oppose the break-up proposal. It also threatened the promoters reform by saying that it would avoid cooperation such as offering special arrangements on taxation and stock market listing.

Consequently the conflict turned into a battle between MPT and MOF. The struggle had reached no conclusion by the end of March, and it was decided to postpone the discussion to 1995. This inconclusive finale partly reflected another feature of the policy network compared with the first case: the absence of powerful policy brokers like Kakuei Tanaka. Tanaka had retired from the political arena owing to illness in 1985. His successors were not powerful enough to settle the debate. This seems to be one of the main reasons why the policy discussion ended up with an unseemingly knockabout debate at the last moment.

The third stage, 1995–1997

The third stage, launched in April 1995, started when MPT once again consulted the Telecommunications Council. The two major cliques referred to in the previous discussion – the NTT clique (NTT, NTT 'family' companies and Zendentsu) and the MPT clique (the 'iron triangle' of MPT, *Teishin-zoku* and NCCs) – were still present. However, there were several changes in their relational structure.

As for the NTT clique, it seemed to have increased internal cohesion since the last discussion. The economic slump after the bubble economy increased the dependence of NTT family companies and Zendentsu on NTT. Telecommunications was one of few promising industrial sectors at that time, and the suppliers owed more of their business to NTT while other markets were diminished. The economic slump also increased anxiety about corporate restructuring and unemployment, which made the trade union more compliant with the management.

Moreover, the end of the LDP's dominance in 1993 changed several conditions under which the telecommunications policy network had functioned. The LDP had been the government party since 1955, but it fell in the public regard after the economic collapse. Fifty-four Diet members left LDP to create a new party in 1993, and formed a coalition government with the LDP. LDP returned to government in 1994, but this time with the help of the Socialist Party, which was now renamed the Social Democratic Party (SDP). Under the traditional system, Keidanren concentrated its support on LDP on the grounds that the party always controlled the government. After LDP left the government, therefore, Keidanren reviewed its strategy and stopped collective donation to LDP, although saying publicly that it aimed to get out of political corruption. The discontinuation of the donation apparently reduced the influence over LDP by NTT and many 'NTT family' companies as important members of Keidanren.

Instead of NTT and many 'NTT family' companies, however, Zendentsu provided a channel of influence from the NTT clique over the government. As already mentioned, LDP formed a coalition government with SDP. Through SDP, therefore, Zendentsu was able to affect the decision-making process of LDP and the coalition government.

With regard to the MPT clique, *Teishin-zoku* seemed to have gained more influence within LDP, since the organized votes by franchised postmasters became more important. This is largely due to the decline of the credibility of the LDP. Furthermore, the voting system was changed from a multiple-seat constituency into a single-seat constituency in 1994. The reform made it even more difficult to have a prospect of election. As a result, the party became more dependent on such stable votes, as organized by franchised postmasters. This gave a greater voice to *Teishin-zoku* within the party.

Another link in the MPT clique, the link between MPT and NCCs, also seemed to have increased mutual connection through *amakudari*. The number of NCCs exceeded 100 in 1995, of which 33 operators offered their executive posts to 52 MPT retirees in total.[25] It is true that *amakudari* executives would not always act on behalf of MPT. The best counterevidence was the employment of MPT retirees by NTT. In the case of NCCs, however, NCCs were apparently quite dependent on MPT, and the offer of executive positions was seen as a return for the regulator's favour.

Apart from those two cliques, MOF had not been very active at the beginning, and did not support either of the two sides. Because of various changes in the telecommunications sector by that time, it became difficult to judge whether the break-up of NTT would have a positive or a negative effect on the asset value of NTT. On top of that, the ministry was not able to pay much attention to the NTT issue, owing to bad loans and other problems that started to surface – most notably '*Jusen*' housing loan companies. In consequence, the reform discussion at this stage was basically made by the first two cliques.

Instead of MOF, several remarkable stakeholders participated in the telecommunications policy network. The first was the Administrative Reform Commission (*Gyokakui*). *Gyokakui* was a special government commission with the mission of administrative reform. It was very similar to *Rincho* in that sense, but unlike *Rincho*, *Gyokakui* was put directly under the supervision of the Prime Minister, with a view to avoiding any interference by narrow ministerial interests. It also had the power to ask the Prime Minister to give its recommendations to particular ministries. Despite such arrangements, however, *Gyokakui* did not seem to enjoy as much political support as *Rincho*. Whereas *Rincho* had strong patrons, Nakasone and Tanaka, that was not the case for *Gyokakui*. Even though *Gyokaui* had direct contact with the Prime Minister, the Prime Minister himself was not very strong at that time. After all, the government was a coalition of three different parties instead of one dominant party, and the Prime Minister, Tomiichi Murayama, was the leader of a minor party, SDP. Hence it was difficult to expect the Prime Minister to give strong leadership in the cabinet.

Gyokakui was different from *Rincho* also in the sense that it focused on deregulation instead of privatization, reflecting growing public scepticism towards control by elite bureaucrats in central ministries. In order to ward off the influence of central ministries, *Gyokakui*'s deregulation subcommittee was mostly composed of people without any bureaucratic career. Consequently, the subcommittee proposed numerous recommendations concerning deregulation, and that was also the case for telecommunications policy. In order to promote market competition in the telecommunications sector, *Gyokakui* put as much emphasis on the deregulation of telecommunications business as on the break-up of NTT.

Gyokakui's position was also supported by another new participant, the Fair Trade Commission (FTC). As the competition authority, FTC's Telecommunications Sector Competition Policy Research Group discussed how the telecommunications market could function effectively. The group had already published its proposal when the break-up of NTT was discussed first in 1989, but it did not seem to exert any influence on the previous discussion. This reflected the fact that FTC was generally regarded as a 'toothless watchdog' in those days.

Yet FTC quickly gained presence in the 1990s. This was not just because Japanese society was more conscious about market competition, but also because the United States encouraged the Japanese government to reinforce FTC primarily as a means to promote its own exports and direct investment. The United States did not seem to have a particular interest in the telecommunications sector, but it was not a good idea for other ministries to downplay FTC, which might have induced direct intervention by the United States. By the time of the discussion on the break-up of NTT, however, FTC had not become powerful enough to overcome MPT. Initially, FTC called for deregulation rather than organizational reform to promote market competition, but later added a statement appreciating the potential benefit of organizational reform, allegedly as a result of pressure from MPT.[26] This was not surprising given FTC's general vulnerability to external pressure.[27] It should be remembered, however, that FTC kept its proposal of deregulation even though MPT did not like it.

In addition to *Gyokakui* and FTC, Keidanren also proposed the break-up of NTT and the deregulation of the telecommunications market simultaneously. There had been a long-standing heated debate at Keidanren, especially between 'NTT family' companies and MPT-retired executives at NCCs. The position of Keidanren fluctuated between the two poles, and finally settled at the middle point.[28]

The emphasis on deregulation as well as the break-up of NTT by those parties transformed the framework of the discussion from a unilateral into a bilateral one. Now it was difficult for MPT to press for the break-up of NTT without making any commitment to deregulation. Such a bilateral framework could promote reform on both sides, but it could also discourage either side from making any progress. The latter turned out to be the case in reality. MPT became reluctant to strongly promote the break-up of NTT because it did not want to promote deregulation. The goal of MPT was to strengthen its position vis-à-vis NTT, and the break-up of NTT was the means for this. Hence, it was paradoxical to commit to deregulation in order to achieve the break-up of NTT. For NTT, it preferred to keep setting the issue aside so as to protect its organizational unity, even though it was interested to promote deregulation.

While *Gyokakui*, FTC and Keidanren collectively made a great contribution to the set-up of the agenda, they were not powerful enough to lead the discussion. Unlike *Rincho*, they were not able to obtain strong political support. This was partly because the government was occupied predominantly with financial problems, and partly because the position of the government on this issue was divided. In the government, LDP often listened to the voice of MPT through *Teishin-zoku*, while SDP tended to support NTT on behalf of Zendentsu.

Accordingly, the coalition government did not include telecommunications policy in its comprehensive deregulation programme in March 1996. The decision was postponed until a year later. However, this conclusion was different in character from the postponement in 1990. In the present case, there was a politician who was able to act as a broker between the two sides – Hiromu Nonaka, the then deputy secretary-general of LDP. Nonaka was the leader of *Teishin-zoku*, but was in a position to take into account wider considerations than Kakuei Tanaka, although not as strongly as him. As the largest coalition partner, LDP might be able to push its position, but Nonaka did not choose to do so, presumably for the sake of the coming election.[29] The best way to keep cooperation in the coalition government was not to make a winner and a loser among the coalition partners.

The reform discussion started again just after the election in October 1996. The result of the election caused a significant change in the structure of the policy network. While LDP steadily increased its seats in the House of Representatives (from 223 to 239), SDP suffered a historic defeat (its number of seats falling from 70 to 15). SDP remained in the government, but LDP was clearly dominant in the coalition. Against this background, Zendentsu accessed LDP to establish a new link, articulating that it was prepared to give support to those who worked on its behalf.[30]

The change of the power balance in the coalition government also gave a condition for a new solution: corporate restructuring by use of a holding company. Applying the framework of a holding company, NTT would be able to preserve substantial unity even though it was formally broken up into several companies. Japanese competition law had traditionally prohibited the establishment of a holding company in order to prevent the concentration of economic power. By mid-1990s, however, the holding company had more often been viewed as a useful measure of corporate restructuring, and demands for the removal of the prohibition had grown. The coalition partner, SDP, initially opposed the reform proposal, as the party was generally cautious about deregulation of the Antimonopoly Law. Nevertheless, it reversed its position after the election. This was primarily because the usefulness of the holding company as the solution for the NTT problem had been widely recognized by that time. Yet SDP might not have prioritized the political benefit of keeping a good relationship with Zendentsu over other considerations if it had not lost so many seats in the Diet.

It was not clear whether the application of the holding company was the best solution for the development of telecommunications in Japan. However, it was politically the best result for both NTT and MPT. NTT won substantial unity. It also gained the possibility of extending into the international telecommunications market, which was now permitted to use its long-distance operation subsidiary. While MPT could not achieve the complete break-up of NTT, it gained a stepping stone to further development.

Although MOF apparently did not join in the discussion, the ministry was certainly satisfied with the result, too. This was shown by the fact that it quickly made several special tax arrangements for the new holding company. What satisfied MOF was not the restructuring of NTT but the reform of the competition law. Indeed, the financial sector seemed to benefit more from the application of the holding company than any other industrial sector.

The fourth stage, 1999–2003

In 1999, NTT was restructured into two regional operation subsidiaries (NTT East and NTT West) and a long-distance operation subsidiary (NTT

Communications) under the holding company. The former subsidiaries of NTT, such as NTT DoCoMo and NTT Data, were also put under the holding company. It is true that a second organizational reform has been argued for, both by those who envisage more independence for the subsidiaries and by those (notably NTT itself) who demand reunification. Nonetheless, no further reform has been accomplished at the time of writing.

Besides the organizational reform, the main focuses in the telecommunications sector have been connection fees and regulatory reform. One of the most remarkable discussions on connection fees was initiated by the demand for lower fees from the United States in 2000. Under the framework of Japan–US deregulation talks, the US Trade Representative (USTR) demanded a cut in NTT connection fees of nearly 50 per cent within a year. In response to that demand, MPT proposed a 22.5 per cent cut over four years. The gap between the USTR and MPT was so wide that they could not settle the discussion by the initial deadline, the Japan–US summit meeting of that year. In the end, MPT promised a 22.5 per cent cut over three years, with a 20 per cent cut in the first two years.

Another important discussion regarding connection fees was whether NTT East and NTT West should retain the standardized connection fees. That was stipulated when NTT was reorganized into those subsidiaries, but only as an interim measure for the three years following the reorganization. Since NTT East and NTT West are formally independent, they have no obligation in principle to set the same fees. On the other hand, those who still see NTT as a united entity, in particular NTT itself, wanted to keep the fees the same. The Telecommunications Council took the former position when it published its recommendation in 2002. Nonetheless, the Diet rejected the recommendation and decided to retain the same fees at NTT East and NTT West.

The US demand for lower fees and the debate on the standardized fees are two different issues, but they commonly indicated a significant change in the structure of the telecommunications policy network: the alliance between NTT and MPT. In the Japan–US talks, MPT acted as the agent of NTT all the time. This was also backed up by MPT's political partner, *Teishin-zoku*. A symbolic episode is that Nonaka, the top figure in *Teishin-zoku* at that time, criticized the United States for its coercive demand, saying that Japan seemed to be being treated as just another US state.[31] Whether he intended it to or not, this encouraged nationalistic resistance while shifting the public focus away from the adequacy of the current fees. However, it should be noted that a group of leading Japanese businessmen also demanded quick achievement of lower connection fees when they talked with US business representatives while MPT and the USTR were discussing the matter.[32]

With regard to the standardized fees, it is true that the Telecommunications Council made a recommendation against the interests of NTT. If the

Telecommunications Council had reflected the interests of MPT, as was usually the case, this should have been regarded as a conflict between NTT and MPT. Nonetheless, the Telecommunications Council functioned differently in that particular case. There was a Council member who firmly argued for the abolition of the universal fees, even though the Ministry of Internal Affairs and Communications (MIC, MPT's successor after the administrative reform in January 2001) disagreed with him. According to this member, Professor Satoshi Daigo, MIC persistently tried to override his argument.[33] In effect, therefore, MIC worked on behalf of NTT in this case too. MIC's efforts failed and a recommendation against the standardized fees was submitted. The Diet eventually rejected the recommendation, but MIC was seemingly much offended by the defiance of the Council. In fact, Professor Daigo was removed from the membership of the Council afterwards, for unclear reasons. Instead, MIC appointed two new members who had once worked for NTT. By that time, it was clear that the MIC–NTT alliance remained central to the telecommunications policy network.

The emergence of the MIC–NTT alliance has sacrificed the traditional relationship between MIC and NCCs. This was illuminated by the discussion on connection fees in 2003, when MIC changed the calculation system for the connection fees apparently with a view to raising the connection fees, on behalf of NTT.[34] NCCs were also critical about the universal fees and some other administrative arrangements. Consequently, five major NCCs brought a suit against MIC, which was quite uncommon in Japan. The *amakudari* connection no longer functioned to secure the dependence of NCCs on the ministry. After all, KDDI Group, whose president had previously headed MPT, now took the initiative in the suit.

Besides the discussions on connection fees, the Telecommunications Business Law and the NTT Law were reformed in 2001 and 2003. The 2001 reform extended the scope of business by NTT East and West, while NCCs were forced to share the responsibility for the contribution to the universal service. The 2001 reform also introduced a system of imposing tougher regulation on dominant operators in order to promote market competition. For the mobile phone market, however, the criterion for market dominance was so low (25 per cent) that it captured not only the NTT subsidiary (NTT DoCoMo) but also other, less dominant operators, which substantially reduced the effect of the regulation. The criterion in the initial proposal was 50 per cent. The initial proposal also suggested the assessment of market competition after two years in order to judge whether a more substantial break-up of NTT should be conducted. Nevertheless, those proposals were deleted because of the opposition of NTT. While the president of NTT was satisfied with the final output (Nikkei communication, 5 April 2001), the MIC official in charge of the proposal was allegedly relegated to a less important position because his initial proposal had offended NTT.[35]

The reform of the Telecommunications Business Law in 2003 changed the regulatory framework, a change that may promote new entry and market competition. However, the operation of the new framework still owed much to the management of MIC. Furthermore, the 2003 reform included the reform of the NTT Law to introduce a new system whereby NTT East subsidizes NTT West to maintain standardized connection fees, which were difficult to justify with the logic of competition policy. Needless to say, this reflected the interest of NTT to reunify its subsidiaries.

Then, what has made MPT/MIC so sympathetic to NTT, despite their antagonism in the previous stages? One explanation may be that the ministry has come to see the preservation of NTT as a single entity to be more beneficial than its break-up. This is not just to retain international competitiveness against foreign companies as the national flagship, but also to facilitate the preservation of regulatory control by the ministry. The dominant position of NTT allows the ministry to intervene in the telecommunications sector. In other words, if the telecommunications market becomes really competitive, MIC will lose its job.

Another explanation for the change in the ministry's attitude may be that it is no longer powerful enough to compete with NTT. There are three reasons for that. First, it is inevitable that deregulation has reduced the scope of government intervention, the main power resource for the ministry. Second, as stated earlier, MPT was succeeded by MIC during the administrative reform in 2001. The post and telecommunications officials were happy to escape from the acquisition of the Ministry of Economy, Trade and Industry, but they were incorporated into a much larger organization, together with other ministries. Now the minister does not work just for the interests of post and telecommunications officials, but also considers other interests. This was especially clear from 2001 to 2003, when the minister was Toranosuke Katayama, the former official at the internal affairs part of the ministry, who had little sympathy for post and telecommunications officials.

The third reason is the emergence of reform of the postal system. This quickly gained momentum after Junichiro Koizumi, a strong advocate of reform, became Prime Minister. Its political supporters, *Teishin-zoku*, are now pre-occupied with that issue and have no time to work for telecommunications. The Prime Minister and his supporters will not be sympathetic towards the telecommunications officials as long as their postal colleagues resist his idea. It should also be remembered that MPT/MIC were strongly opposed to the break-up of the postal service, whereas the ministry had long argued for the break-up of NTT in the previous stages. Since the ministry and *Teishin-zoku* benefited from the national franchise of the postal service, they wanted to preserve it. Against this background, it is not surprising that they stopped talking about the break-up of NTT when others started to discuss the break-up of the postal service.

The entry of new operators also seems to have affected the attitude of the ministry to NTT. Foreign-owned operators, such as Vodafone, may utilize their home country to impose political pressure in the form of trade negotiations. Even some domestic operators are now quite defiant of the traditional way of regulation. Softbank is a case in point. The company does not hesitate to break the unwritten rules that NTT and the other older NCCs have long followed unconditionally.[36] Apparently stimulated by Softbank, other NCCs have also become more ambitious than before, as shown in the case of the administrative suit in 2003. In those circumstances, NTT may look a better partner for the ministry even though it is not very subordinate.

The emergence of the MPT/MIC–NTT alliance has transformed the previous two-clique structure of the telecommunications policy network. A big clique has emerged, encompassing the ministry, NTT, *Teishin-zoku* and the NTT union (formerly Zendentsu), with other interests being marginalized.

Another change in the telecommunications policy network from the previous stage was the absence of powerful policy brokers like Tanaka and Nonaka. Nonaka had been a Diet member until he retired in 2003, but he had already lost his central position in LDP when his rival, Koizumi, became Prime Minister. Another important political stakeholder, SDP, also became powerless in the network. The party became very weak after leaving the coalition government, and Zendentsu shifted its support to the new second-largest party, the Democratic Party (DP). However, the party is on the whole reluctant to act straightforwardly for the sake of its supporters, as it often criticizes political patronage by LDP. In this respect, the case of the supplementary resolution accompanying the 2003 reform was noteworthy. When the House of Councillors passed the reform bill, it attached the resolution that NTT's obligation to open its fibre-optic network should be reconsidered. The resolution was proposed by two DP members who had once worked for NTT and received much electoral support from the NTT Union. Nevertheless, they were not in leading positions themselves, nor did the party leaders support them. As a result, they faced strong criticism within the party. The resolution was not supported by LDP either, and thus it was struck down by the House of Representatives. While the reasonability of the resolution was debatable, it was at least clear that the absence of powerful policy brokers contributed to such a less orderly process.

Conclusion

The present study traces the history of the policy-making process for Japanese telecommunications policy for the past 20 years from the perspective of policy network. Each case has a different focus, but all cases support the view that the policy outputs do not rest purely on technological grounds, as stated at the beginning of the study. Instead they often depend on how the stakeholders make alliances and interact with each other. The cases also show that the structure of the policy network is far from constant. Therefore, it is difficult to predict future results of the policy-making process. The absence of powerful policy brokers also seems to have reduced predictability. At least, however, it is clear that not only private operators but also the ministry officials and the semi-public NTT have often acted to maximize their own interest, instead of only considering the public interest.

When it comes to the centrality of the policy network, it is remarkable that NTT has moved more to the centre of the network. This may be counterbalanced by the reduction of the company's economic influence by the entry of newcomers, resulting from such factors as technological development and market globalization. At this moment, however, it seems unlikely that NTT will lose its economic power so radically as to lose its political power too and move out of the centre of the telecommunications policy network.

Against this background, it is questionable whether MIC can achieve a good performance as the supervisor of the telecommunications market. After all, no competition regulator will be successful if it takes account of interests other than the promoting of fair competition. It is therefore not surprising to envisage FTC playing a better role. FTC is able to regulate unfair trade practices even under the existing framework, but it always suffers from criticism concerning 'dual regulation' when it takes action. In turn, FTC once proposed the inclusion of the telecommunications sector within the ordinary competition policy,[37] but the proposal did not draw much attention from the public. The lack of public attention was evident when few argued against the merger of FTC and the postal and telecommunications officials into the same ministry (MIC) after the administrative restructuring. FTC was moved from MIC to the Cabinet Office in 2003, but this was largely the consequence of demands from the United States. Independence is an important strength of FTC, but it is often translated into isolation, hence its weakness vis-à-vis those parties with strong political resources. This seems to be one of the reasons why NTT (NTT East) dared to reject the recommendation of FTC to modify its pricing policy for fibre-optic services and openly criticized FTC's argument.[38] FTC dropped its proposal of reinforcing its monopoly/oligopoly regulation to open the use of essential facilities, including the telecommunications network to newcomers when it discussed the reform of the Antimonopoly Law in 2004, allegedly as a result of the opposition of politicians and other ministries, including MIC.[39]

While FTC is largely free from political intervention, it often lacks political support for the development of competition policy, as shown in the above reform case. Not surprisingly, few Japanese politicians, particularly LDP members, are willing to support competition policy, which often collides with the interests of big businesses, many of which are their sponsors. As a result, it is often difficult for FTC to act powerfully, unless competition policy attracts much public attention, as in the mid-1970s, or the government is under strong external pressure, as in the early 1990s. Furthermore, FTC does not enjoy so much credit in general. For instance, a university professor publicly criticized NCCs not so many years ago when he asked FTC to investigate NTT's practices, saying, 'companies asking FTC for such a thing are cowards [*darashi-nai*]'.[40] However, the professor has been a member of the Telecommunications Council and he might have made such a criticism intentionally, in favour of MIC.

Those problems must be addressed if a new supervisory body were to be established. Whereas a new body should be independent of the control of MIC, it may not be successful without substantial public support.

Apart from the institutional choice, it may be necessary to reconsider the operation of competition policy itself. The cases that have been discussed demonstrate the significance of the political power of NTT in the performance of the telecommunications market, but the traditional approach of competition policy only takes economic power into consideration. To discuss how to incorporate political factors into the assessment of market power is beyond the scope of the present chapter, but a possibility to develop competition policy in that direction should be considered, at least in the current context of the telecommunications sector in Japan.[41]

Notes

I acknowledge the support of Ms Miki Futagawa, EIJS Tokyo Office, for collecting relevant data in Japan. I am also thankful to Professor Masako Wakui for providing valuable materials.

1 M. Muramatsu, 'Mineika, Kiseikanwa to Saikisei no Kouzou – Denkitsuhinseisaku no Henka' (The Structure of Privatization, Deregulation, and Reregulation: Changes in Telecommunications Policy), *Leviathan* 2: 118–135 (1988); C. Johnson, 'MITI, MPT, and the Telecom Wars: How Japan Makes Policy for High Technology', in C. Johnson and L. D. Tyson (eds) *Politics and Productivity: The Real Story of Why Japan Works* (Cambridge, MA: Ballinger, 1989, pp. 177–240); I. Gow, 'Re-regulation, Competition and New Industries in Japanese Telecommunications', in S. Wilks and M. Wright (eds) *The Promotion and Regulation of Industry in Japan* (New York: St Martin's Press. 1991), pp. 256–285); R. G. Noll and F. M. Rosenbluth, 'Telecommunications Policy: Structure, Process, Outcomes', in P. F. Cowhey and M. D. McCubbins (eds) *Structure and Policy in Japan and the United States* (Cambridge: Cambridge University Press, 1995, pp. 119–176); S. K. Vogel, *Freer Markets, More Rules* (Ithaca, NY: Cornell University Press, 1996); E. Kawabata, 'Sanction Power, Jurisdiction, and Economic Policy-making: Explaining Contemporary Telecommunications Policy in Japan', *Governance: An International Journal of Policy and Administration* 14 (4): 399–427 (2001); M. Tilton, 'Ideas, Institutions, and Interests in the Shaping of Telecommunications Reform: Japan and the US', in L. Weiss (ed.) *States in the Global Economy: Bringing Domestic Institutions Back In* (Cambridge: Cambridge University Press, 2003, pp. 180–199); M. Tsuchiya, '1990-nendai no Joho Tsushin Seisaku – NTT Keieikeitaimondai ni

Torawareta Juunen' ('Japanese Telecommunications Policy in 1990s: The Debate on NTT's Breakup'), *Leviathan* 32: 72–96 (2003); I. Toyonaga, 'Seikai-saihenki ni okeru Koporatisuto Sinario no Taitou to Zasetsu – NTT Bunkatsuronsou no Kisuu wo meguru Ichishiron' (The Rise and Fall of a Corporatist Scenario: Politics over the Break-up of NTT in the Coalition Period), *Leviathan*, 32: 97–122 (2003).

2 R. A. W. Rhodes and D. Marsh, 'Policy Networks in British Politics: A Critique of Existing Approaches', in D. Marsh and R. A. W Rhodes. (eds) *Policy Networks in British Politics: A Critique of Existing Approaches* (Oxford: Clarendon Press, 1992, p. 3).

3 M. M. Atkinson and W. D. Coleman, 'Policy Networks, Policy Communities and the Problems of Governance', *Governance: An International Journal of Policy and Administration*, 5 (2): 154–180 (1992).

4 A. Cawson, 'Introduction: Varieties of Corporatism: The Importance of the Meso-level of Interest Intermediation', in A. Cawson (ed.) *Organised Interests and the State: Studies in Meso-corporatism* (London: Sage, 1985, pp. 1–21).

5 Muramatsu, 'Mineika, Kiseikhanwa to Saikisei no Kouzou'; Toyonaga, 'Seikai-saihanki ni okeru Koporatisuto Sinario no Taitou to Zasetsu'.

6 G. Jordan and K. Schubert, 'A Preliminary Ordering of Policy Network Labels', *European Journal of Political Research* 21: 10 (1992).

7 Rhodes and Marsh, 'Policy Networks'; F. Van Waarden, 'Dimensions and Types of Policy Networks', *European Journal of Political Research* 24: 29–52 (1992).

8 See, for example, J. Scott, *Social Network Analysis: A Handbook* (London: Sage, 1991).

9 P. A. Sabatier, 'An Advocacy Coalition Framework of Policy Change and the Role of Policy-Oriented Learning Therein', *Policy Sciences* 21: 129–168 (1988).

10 Vogel, *Freer Markets, More Rules*, p. 141; Tsuchiya, '1990-nendai no Joho Tsushin Seisaku'.

11 Y. Suzuki, *Anto! NTT vs Yuseisho* (Struggle under cover! NTT vs Ministry of Post and Telecommunications) (Tokyo: Kodansha, 1996, pp. 50–54).

12 Ibid., p. 54.

13 Arinobu Moriya, the Cabinet/Communication Committee, 2 July 1982.

14 Vogel, *Free Markets, More Rules*, pp. 153–154.

15 See the debate between Jinichi Katayama and Arinobu Moriya, third section of the Budget Committee, House of Councillors, 27 March 1981.

16 T. Inoguchi and T. Iwai, *Zoku Giinno Kenkyu* (A Study of Zoku Politicians) (Tokyo: Nihon Keizai Shinbunsha, 1987).

17 Nihon Keizai Shinbunsha, *Gekishin Dokyumento NTT* (Great Shock – Document NTT) (Tokyo: Nihon Keizai Shinbunsha, 1989, p. 42).

18 Vogel, *Free Markets, More Rules*, p. 152.

19 Ibid.

20 Ibid., p. 150.

21 Ibid., p. 154.

22 Nihon Rodo Kenkyu Kiko, *Denki Tsushin Sangyo no Roshi Kankei – Rekishi to Genjo* (Industrial Relations in the Telecommunications Sector: History and Current Situation) (Tokyo: Nihon Rodo Kenkyu Kiko, 1996, p. 67).

23 F. Fujii, *Bunkatsu – NTT vs Yuseisho* (Breakup: NTT vs Ministry of Post and Telecommunications) (Tokyo: Daiyamondosha, 1996, p. 88).

24 *Nihon Keizai Shinbun*, 9 March 1990.

25 *Yomiuri Shinbun*, 4 November 1995.

26 *Nihon Keizai Shinbun*, 20 November 1995.

27 K. Suzuki, *Competition Law Reform in Britain and Japan: Comparative Analysis of Policy Networks* (London: Routledge, 2003, pp. 147–152).

28 Fujii, *Bunkatsu*, pp. 67–86.
29 *Nikkan Kogyo Shinbun*, 29 March 1996.
30 *Nikkan Kogyo Shinbun*, 26 November 1996.
31 *Sankei Shinbun*, 26 August 2000.
32 *Nikkan Kogyo Shinbun*, 12 July 2000.
33 S. Daigo, '*Soumushoga shinguikai-iinno watashiwo "kainin" shita shinsou*' (The real reason why the Ministry of Internal Affairs and Communications 'dismissed' me from Council membership), *Ekonomisuto*, 21 January: 73–75 (2003).
34 The NTT connection fees are calculated basically by dividing the cost by the amount of the traffic in a reference year. When MPT decided the fees for 2000–2002, it referred to the traffic in 1998 instead of using the current years. Since traffic was on the increase at that time, the connection fees were higher than they would have been if the current year had been taken as reference point. When the ministry revised the fees for the years 2003–2004, however, it suddenly changed the system so that the connection fees would reflect current traffic. The traffic was declining at that time, so the connection fees would very likely be higher than they should have been with reference to the past year.
35 *Nihon Kogyo Shinbun*, 2 July 2001.
36 Nikkei Communication, *Shirarezaru tsushin senso no shinjitsu: NTT, Softbank no anto* (The Unknown Truth of the Communication War: NTT – Softbank Struggle under Cover) (Tokyo: Nikkei BP, 2003, pp. 15–23).
37 Fair Trade Commission, *Koeki jigyo bunya ni okeru kisei kanwa to kyoso seisaku* (Deregulation and Competition Policy in the Public Utility Sectors) (Tokyo: Fair Trade Commission, 2001).
38 *Nikkei communication*, 25 February 2004.
39 *Sankei Shinbun*, 27 May 2004.
40 *Mainichi Shinbun*, 25 October 2000.
41 With regard to this, Machida provides a detailed analysis as to how NTT has utilized its political power to enhance its market power. T. Machida *Kyodai dokusen: NTT no shukuzai* (Giant Monopoly: NTT's Remaining Wrongdoing) (Tokyo: Shinchosha, 2004).

8 Ill-defined national interest

The difficult role of the Japanese negotiator in the access charge negotiations with the United States

Motohiro Tsuchiya

Introduction

The definition of diplomacy is to adjust interests between states by any means. However, a precondition of such diplomacy is for a state to make clear its own national interests. If not, what would the outcomes of a negotiation be?

There was an example of such negotiations between Japan and the United States, in which the Ministry of Internal Affairs and Communications (MIC) could not easily find points of agreement regarding access charges within the Japanese market. Access charge is a charge that is settled between a principal electronic telecommunications operator that retains essential facilities to a dominant degree and a company that is competing in the market and trying to rent the facilities essential for its own business.

The issue of access charge is not necessarily argued in international negotiations. However, the US administration requested the Japanese government to relax the regulations in the telecommunications industry and put this issue on the agenda in bilateral talks. Therefore, while MIC held talks with the United States Trade Representative (USTR), it was compelled to compromise regarding the interests between Nippon Telegraph and Telephone (NTT), the former dominant operator holding the Japanese local loop, and other competing operators in the market.

In this chapter, we examine the chronological transition of the access charge issues during the three years 2002–2004 and the roles the MIC was expected to play at the time of the transitional phase in the telecommunications industry. Then an analysis is made of the relationship between the market competition policies and national interests in international negotiations. MIC holds two roles: it is a regulator of the market and a negotiator with foreign counterparts. However, these two roles have produced conflicting and difficult situations for MIC, which is otherwise supposed to encourage the competing operators to reach compromises, define national interests and take tough measures in negotiations without any obstacles.

Two-level game

Interaction of domestic politics and international politics

There have been arguments about the existence of some forms of relationships between domestic politics and international politics. James N. Rosenau called such relationships 'linkage politics'.[1] Karl W. Deutsch and Ernst Haas analysed the regional integration of Europe and studied the results found between domestic development and international development.[2] Robert Keohane and Joseph S. Nye also pointed out the roles that domestic actors play, to the extent that they can influence international relationships within a theoretical framework of interdependence.[3]

In his book *Essence of Decision*, which analysed the Cuban crisis, Graham Allison suggested a bureaucratic political model as well as an organizational and rational selection model. He pointed out the roles domestic concerns can play influencing decisions in the international arena.[4] Peter Katzenstein and Stephen Krasner developed arguments about structural state power and defined the 'strength of the state' as a capacity of the government to be able to contain domestic actors.[5] For example, the United States is a 'weak' state, owing to strong domestic lobbying activities, while Japan is a 'strong' state, since the government can contain the national lobbying groups and other interested groups. However, how the strength of such containment is measured in real terms is left unanswered.

Robert Putnam of Harvard University analysed the interaction of domestic and international politics by using the 1978 G7 Summit as an example.[6] He called the analytical framework a 'Two-Level Game' and made the assumption that a state in international negotiations cannot be regarded as a single actor. In other words, although we normally personify the state, such as 'Japan is...' or 'the United States does...', as if it is a sole actor, no such abstract description can be established in reality. This is because a state is a compound entity consisting of many participating and interacting actors, and in no circumstances do such actors unanimously make the same choices.

Therefore, Putnam established two levels of analysis: a domestic level and an international level. Level 1 is a phase where negotiations between governments are held with a view to reaching an interim agreement; Level 2 is a phase where arguments are conducted by domestically powerful or supporting actors to determine whether the agreement can be ratified. Since arguments at Level 2 can influence or even damage agreements reached at Level 1, the necessity at Level 2 to ratify the agreements can influence the agreement process at Level 1. Therefore, governments can experience difficulties in reaching agreement after the negotiations at Level 1, despite their intentions to finalize the deal.

At the domestic level, certain domestic groups pursue their own interests by lobbying governments to adopt favourable policies for themselves while the politicians pursue power by forming an alliance among such interested domestic groups. At the international level, many governments try to minimize the intervention of foreign powers by attempting to maximize capacity to satisfy domestic lobbying and interested groups. However, policymakers in the central government cannot ignore both domestic and international powers when negotiating with a foreign state. A political leader in each state is like someone seated ready to play a board game at both domestic and international levels.

The metaphor of two tables can explain what cannot be explained by using only a single-actor model. For example, a certain movement that a player regards as rational on one table may become an irrational choice for the same player on the other table. Some powerful and influential groups may overturn the table if they do not like the results. In addition, a tiny movement on one table may trigger regrouping on the other table. Although this model is useful to analyse the interaction of domestic and international politics, it should be noted that problems of the perception gap in negotiations remain unsolved. Four possible risks for a negotiator may be listed a negotiator

1 misunderstands Level 2;
2 misunderstands counterparts' domestic issues and preferences;
3 thinks there is room for information control;
4 is left to judge influence from international pressures.

In other words, it is important for negotiators to clearly judge their own national interests and their counterparts' national interests, compare their own room for negotiations with the counterparts' room for manoeuvring, and gain more advantageous results for their own state from the negotiations. If they can define the national interests precisely and clearly without being disturbed, then they will be a front-runner in negotiations.

On the other hand, if the national interests cannot be defined precisely, a negotiator may find it difficult to reach a compromise. This is the category that the Japan–US deregulation talks on the access charge fall into.

Before we move on to the details of the talks, let us have a look at another cause of the stalemate in the access charge talks. It can be said that LRIC (long-run incremental cost) or TELRIC (total element long-run incremental cost),[7] which is an access charge calculation model, contributed to the negotiation stalemate since its definition was complicated and abstract.

Confused models

One of the most important concepts dealt with in the Japan–US DEREGULATION Talks is the access charge and its calculation model. In the network industry, it is said that connecting a network with another network brings enhancement of externality. By expanding the network, the number of users increases substantially and users will find the service more convenient.

However, building such a wide network comes at a high price. If the networks can be connected at the same scale and cost, connecting them can be dealt with easily. If the networks to be connected are asymmetrical, interconnection fees are needed. Networks in the real world can never be symmetrical. Operators are continuously investing in networks step by step, depending on their business needs. Each network is built on specific costs.

Since no appropriate method has been established concerning the calculation of interconnection fees, however, two approaches, roughly, are taken: a bottom-up approach and a top-down approach. The bottom-up approach is based on virtual costs required to reproduce an existing network by using the most usable latest technology. LRIC is included in this category. On the other hand, the top-down approach, or historic approach, is based on actual costs to establish the network.

In general, access charges are cheaper in the bottom-up approach than in the top-down approach. Therefore, a newly entering operator favours the bottom-up approach, although the existing operators favour the top-down approach so that they can recoup the money they invested in the network. There is no established theory as to which is an appropriate or better approach. It can be said therefore that each state is using either approach according to its policy goal while it takes both approaches into consideration.

However, the problem is that each state is negotiating without establishing internationally standardized definitions. There are three types of confused areas.

First, the United States and the United Kingdom use different definitions of 'interconnection' and 'access charge'.[8] In the United States, interconnection means only the physical linking of networks and does not include the cost of transporting traffic. In addition, there are two meanings of access charge: (a) the cost to deliver the traffic to the users; and (b) supplements of the deficits against access cost.

On the other hand, in the United Kingdom, interconnection includes (a) from the definition of access charges as defined above in addition to the same meaning of physical linking as defined in the United States. When it comes to access charges in the United Kingdom, they are limited to (b) as regarded in the United States.

Second, the meaning of terminology used in the Japan–US talks is not consistent. In Japan, interconnection means the same access charges as defined in the United States. The Japanese mass media do not use the terminology of access charges, and the whole concept is simply represented in the term 'interconnection' (*sougo-setsuzoku*).

Third, although the US administration is requesting the Japanese government to introduce LRIC as the access charge calculation model, the common definition of LRIC is not shared between both countries. The US administration defines it within a different context;[9] in the United States, LRIC is used for interconnection between local telecommunications operators but not used in the event that a long-distance call operator accesses local loops, which a local operator controls. On the contrary, in Japan the applicable sphere of LRIC is understood differently as compared with the United States: access charges are applicable to all operators that have access to local loops owned by local operators such as NTT East and NTT West.

LRIC is not an established model and each country has been using slightly different models. The UK government determines access charges by taking into consideration various factors found in both the top-down approach that British Telecom utilizes and the simulated bottom-up approach. This type of confusing terminology definition has made it difficult to find a resolution of any agreements between Japan and the United States.

Political settings

What would be the outcome if we were to apply the two-level game framework to the Japan–US talks on access charges? As examined above, access charges are subject to interpretation. In this chapter, access charge in the Japanese market has two meanings. First, access charges are to be paid by operators, including NTT Communications, a long-distance carrier and Internet service provider under NTT Holding Company, in order to access the local loops owned by NTT East and NTT West. The local loops owned by NTT East and West are an essential facility and needed by other operators so that they can deliver their own services to their own customers. In other words, access charges are a cost paid by these operators to use the essential facilities, instead of building their own facilities or systems.

Second, access charges mean the amount of charges that a mobile phone operator such as NTT DoCoMo, 'au' and TU-KA of KDDI Group, or Vodafone (ex-J Phone) levies from fixed-line telephone operators and international call operators. Mobile phone operators keep down the charges that their own customers have to bear (from a mobile to a fixed line), while they levy high access charges (from a fixed line to a mobile), which their customers have no obligation to pay.

Since the mobile phone operators were given freedom to set the charge price for a fixed-line to mobile access (or, we could say, 'mobile termination fee') until recently, this was regarded as a main source of revenue. However, it is expensive, for example, when a person in the United States calls another person on his or her mobile in Japan. Therefore, the USTR began to request Japan to lower the access charges.

However, a big change occurred when a new service called '*Chokushu*

Service' was introduced by Softbank. This picks up lines directly from residences by connecting them to facilities owned by New Common Carriers (NCCs) inside NTT's bureau offices. Having purchased Japan Telecom in July 2004, Softbank acquired wider access to the dry copper of NTT. On 30 August, Softbank chairman Masayoshi Son announced the commencement of a new fixed-line telephone service from December 2004, the basic charge of which comes cheaper by around ¥200 than that of NTT.

In the next section, the progress of the Japan–US talks is analysed. When the telecommunications policies are amended, patterns as mentioned below are likely to be repeated:

1 production of draft proposal by study groups in MIC;
2 MIC's consultation with the Telecommunications Council;
3 production by MIC of bills and regulations after the reports prepared by the Council;
4 law amendment in the Diet or promulgation of regulations by MIC.

The main goal of the US administration is to acquire freedom of business for US companies within the Japanese market. It was a challenging goal for the US administration and the industries to open up the Japanese market in various fields. While cheap, high-quality Japanese products were storming the US market, they thought that it was unfair that US products or services with genuine competitive power were being excluded from the Japanese market. Therefore, they tried to open up the Japanese market, even through political coercion. It seems that previous US administrations could raise their diplomatic reputations based on the Japanese market actually 'opening if knocked'.

It was a bitter scenario for those who regarded Japan as a partner in terms of national security issues in Asia that economic ones were becoming highly politicized. When no more important issues than economic ones are found, access charges in the telecommunications sector, which are not necessarily important, are treated as an issue in the talks.

However, it is difficult for Japan to define its goal – in other words, national interests – in the negotiations. First, it was the United States that offered talks, and Japan lacked incentives to be actively engaged with them. Second, the powerful and influential groups within Japan are divided: on the one hand we see NTT, which stopped the access charge cut; on the other hand are NCCs, which were still strongly demanding a cut in access charges. Third, there were no active actors interested in the issues behind the US administration. It seems that no actors were interacting to maintain some kind of benefits after the talks. Therefore, it was likely that the USTR was showing a hard-line attitude only to score points in its reputation for diplomatic skills.

The United States could define its national interests clearly by strongly insisting on a cut in NTT's access charges, while Japan had to adjust access charges to 'appropriate levels'. Therefore, it was assumed that this negotiation was a difficult one for Japan. In the next section, the actual Japan–US talks are examined in detail.

Power game in the Japanese market

The 1999 regime

The three-year period this chapter deals with lies between the beginning of 2002 and the end of 2004. However, the beginning of access charge issues goes back to 1985, when the Nippon Telegraph and Telephone Public Corporation (NTTPC) was privatized, with a view to dividing the company. The break-up of AT&T in 1984, based on a modified final judgment in 1982 in the United States, had already occurred. At that time, there were many experts who advocated the break-up of NTTPC in order to bring competition to the telecommunications market. However, in the event, the newly born NTT was not broken up because its privatization was prioritized.

With the birth of NTT, there were also newcomers to the market such as DDI and Japan Telecom, and the telecommunications market in Japan formally took a liberalization step and introduced competition. However, it was NTT that could yield power on the market since it had inherited a large part of its wealth from the NTTPC, and newcomers were forced to compete from an inferior position to NTT on the domestic telecommunications market, although they could progress on the corporate market to some degree.[10] Therefore, the Ministry of Posts and Telecommunication (MPT)[11] was aiming in its policies at stimulating new business entries and introducing competition. To achieve this aim, the break-up of NTT, which was not realized in 1985, became a possibility.

The issue of NTT break-up was mooted in 1990 and again in 1995, but NTT succeeded in preventing it. In December 1996, however, Hiromu Nonaka, secretary-general of the Liberal Democrat Party (LDP), mediated as a broker between the CEO of NTT, Junichiro Miyazu, and the Vice-Minister of Posts and Telecommunications, Mitsuo Igarashi, to bring an end to the stalemate of the NTT problem. In this case, political influence and power settled the issue. Subsequently, in June 1997, the NTT reform bill was passed, and in July 1999 NTT, under the holding company, was restructured into NTT East and NTT West as local telecommunications subsidiaries and NTT Communications as a long-distance call and data communications subsidiary. NTT DoCoMo, NTT Data and many other 'family' companies were reorganized under the holding company.

With the completion of restructuring of NTT in 1999, it seemed that the MIC had obtained a certain achievement in an issue left festering for so long. On the other hand, although NTT faced a slight setback in its management structure, it could still hold the group management under the umbrella of the holding company. The case of NTT was an extremely political solution that incorporated the intentions and demands of both sides concerned.

Under this 1999 regime, NTT East and NTT West, as local telecommunications operators holding essential facilities, came to be placed under strict regulation. Consequently, the charges and other fees required the approval of MIC, and their business was no longer free from MIC's control. As MIC tended to take a pro-NCC stance, as seen in the historical linkage, NTT strived to retain its predominant competitive position on the market, by being strictly defensive in its stance. It also resorted to using the influence and power of political leaders in the national Diet in order to compete against the MIC.

The most important issue after the reform was access charges and market opening issues, as recognized by the United States. In July 2000, the G8 Summit was planned to take place in Okinawa, Kyushu, in southern Japan. The Japanese government as host state intended to settle all the unpleasant political issues before the summit. By contrast, the US administration took this as an opportunity to try to extract a compromise from Japan by openly suggesting, in a rather threatening manner, that it would engage in arguments in front of other leaders at the summit. Furthermore, the US administration insisted that a cut in access charges would stimulate the telecommunications market in Japan and contribute to the expansion of the Internet, the result of which was regarded as in the interest of the Japanese public as a whole.[12]

During the Birmingham summit, held in May 1998 in the United Kingdom, the Japanese and US governments had agreed on certain issues, including access charge. This agreement stipulated that Japan would at the earliest time introduce a new calculation system to lower interconnection fees. An amendment bill on the Telecommunications Business Law to introduce LRIC was submitted to the Diet in spring 2000 before the Okinawa summit. However, the US administration continued to intervene on a regular basis to request a cut in access charges. As a final result, Japan and the United States informally agreed on 18 July 2000 that Japan would cut access charges by 22.5 per cent in three years' time, and that more than 20 per cent, which consists of about 90 per cent of the total deal, was expected to be cut within the first two years. Therefore, MIC requested NTT to reduce its access charges and NTT unwillingly agreed.

In 2001, the service called 'My Line', which set default telecom operators without prefix numbers, was introduced to stimulate competition in the Japanese telecommunications market. Each telecommunications operator launched a large-scale advertising campaign to gain more share of the market. Subsequently, it turned out that NTT-related companies kept their previous dominant position in the market. However, since the introduction of My Line services triggered price reduction competition on the market, it reduced the revenue of NTT and raised uneasiness about the future prospects of the telecommunications market in general. In addition, the Telecommunications Business Law was amended, which obliged NTT to open its fibre-optic network at reasonable prices to other operators if they request to use them. NTT responded to this amendment by stating that the obligation to open its network to others was not appropriate because the infrastructure of NTT's optical fibre was not an inheritance from the NTTPC, but an outcome of competition. For example, Tokyo Electric Power Company has its own fibre-optic network, which can compete with NTT's.[13]

Furthermore, a new service that eventually brought a big historical change commenced in December 2001. Heisei Denden Co. launched a cheap telephone service by using dry copper owned by NTT.[14] Since the service area covered by Heisei Denden Co. was limited, this move did not attract much attention. However, it was a remarkable step in that Heisei Denden Co. provided a milestone in the telecommunications industry for fixed-charge telephone service competition, opening the way for operators such as Softbank.

Declining fixed-line telephone system

It was a study group within MIC that spearheaded the series of discussions to come. On 1 February 2002, a report was presented, stipulating that NTT's access charges could be cut by 19 per cent at a maximum in financial year (FY) 2002.[15] The NTT chairman, Miyazu, contested the report in the regular press conference held on 6 February, saying that the fiscal situations of two companies (NTT East and NTT West) had been worsening and there was no room to discuss the reduction. Similarly, NTT East and NTT West predicted a substantial fall in revenues in a business plan submitted to MIC at the end of the same month. However, MIC's study group completed the report on 9 March, advocating that NTT was capable of cutting the charges by 19 per cent at maximum.

It should be noted that this report was produced in the light of the Japan–US senior administrative negotiations held on 13 and 14 March. In the talks, the United States requested a cut in access charges, although no agreement was reached. On 28 March, MIC consulted with the Telecommunications Council regarding the review of a calculation method of access charges.

On 3 April, the USTR published a report on the 2002 Foreign Trade Barrier, which is based on Article 1377 of the Omnibus Trade and Competitiveness Act.[16] In this report, the USTR requested a cut in access charges to Japan in general and criticized NTT DoCoMo's access charges in particular as being inappropriately high. Responding to this, NTT DoCoMo's chairman, Keiji Tachikawa, argued in the press conference that 'we levy [access charges] fairly'. On 17 April, MIC also submitted the documents that rejected its requests to the USTR.

NTT announced its three-year plan covering the period up to 2004 on 19 April, and it said it would stop in principle the investment in switching devices used for the fixed-line telephone system. That is, NTT was alleged to have admitted that there was no real future in the expansion of the fixed-line telephone system and such a sector would be subject to reduction in scale.

Then, on 14 May, the long-awaited verdict was given at the US Federal Supreme Court. In this case, US local telephone operators regarded the introduction of LRIC to the US market as unfair. Therefore, if the court objected to the introduction of LRIC, it was thought that its decision would have a great impact on the Japan–US talks. However, the court ruled in favour of the introduction of LRIC; the US administration was relieved, while the ruling disappointed the Japanese government. However, the case dealing with the introduction of LRIC itself demonstrated that LRIC had potential to raise doubts about its practicability as an access charge calculation model. In the Japan–US talks held on 20 May following this ruling, the United States again requested a cut in access charges, especially those of NTT DoCoMo.

At that time, the fixed-line telephone services entered into a new period of decline, to a noticeable degree. On 31 July, NTT East and NTT West announced their FY 2001 connection accounts, with a reduction by 3.2 per cent more than the previous year. Next day, the Telecommunications Council, however, recommended that access charges be reduced further by 8 per cent, or 51 per cent, in FY 2003. Although the preconditions that lead to these two calculations differ, the MIC was concerned about an unrealistic reduction of 51 per cent and favoured the 8 per cent plan.

On 28 August, the KDDI chairman, Tadashi Onodeara, announced his concerns over access charges, saying that in the event that the latest communication amount data were used for a calculation, access charges to connect with local switching devices would also go up (the Council's report expected them to fall). In short, access charges are calculated from the formula of the cost/communication amount. However, since the denominator of the communication amount had been decreasing, then as long as the current model was applied, access charges had the potential to go up. Onodera requested a review of the model if the original goal was not attainable: reducing access charges and securing clarity.

In the event that access charges were to be reduced, it was assumed that NTT might make up the loss incurred by the reduction by increasing the basic charge. At the hearing organized by the Council on 29 August, participants from the telecommunications sector and consumer groups exchanged opinions and raised concerns about the transfer of such levying methods to the basic charge. On 4 September, MIC disclosed opinions from the US administration while the latter criticized Japan on the grounds that access charges were too high compared with other competing markets.

During the spring of 2002, MIC started discussing the possibility of approving the gap between NTT East and NTT West. The reason why the MIC considered this option was that the area NTT West covers consists of more islands and islets than that covered by NTT East, and as NTT West tried to satisfy the universal service obligation, it became burdened with greater costs, which had eventually contributed to its internal structural deficits.

Table 8.1 shows the operating earnings and ordinary income of NTT East and West between FY 2000 and 2003. As of May 2002, when the closing account of FY 2001 was announced, the ordinary income of NTT East experienced a huge loss and that of NTT West also demonstrated that the negative rate had been growing compared with the previous year. As far as the figures are concerned, the deficit finance of NTT West was based on a statement. To cover the deficits, it then started being suggested that retaining the same fixed rate between NTT East and NTT West was not necessarily imperative.

On the other hand, activities to request a cut in NTT's access charges were proceeding. While the Council approved the gap between NTT East and West, it completed a report on 13 September, in which it requested a cut in access charges of between 8 per cent and 16 per cent in FY 2003. Furthermore, in the Japan–US talks held in Washington DC on 28 October, the United States again expressed its concerns over access charge issues.

However, this type of approach to approve the gap between NTT East and NTT West was suddenly overturned on 28 November. This was because NTT East and NTT West wanted to retain the close management relationship internally, and they began political lobbying. As a result, the committee that deals with internal affairs and management in the House of Representatives demanded that NTT East and NTT West work to narrow the gap between themselves. This move by the Diet eventually contributed to preventing the gap from widening, a gap that MIC and the Telecommunications Council had already decided to approve by then.

Table 8.1 Operating earnings and ordinary income of NTT East and NTT West (million yen)

Fiscal year	*NTT East*		*NTT West*	
	Operating revenue	*Ordinary income*	*Operating revenue*	*Ordinary income*
FY 2000	2,794,500	14,100	2,639,500	−105,700
FY 2001	2,573,600	7,509	2,406,700	−170,495
FY 2002	2,352,200	63,315	2,215,000	44,925
FY 2003	2,267,800	97,853	2,166,800	90,560

Source: NTT East and NTT West.

Therefore, on 5 December, MIC finalized the policy to set up a new system, by which NTT East was expected to give financial support to NTT West.

Battle over access charge raise

On 18 December 2002, the MIC study group published its perspective that the amount of communication traffic in FY 2004 would be less than one-third of that in FY 2001, owing to traffic transfer from fixed-line telephones to mobile or IP phones. Any reduction in communication traffic could automatically lead to a rise in access charges. In the Japan–US talks held two days after the publication of the report, MIC explained the possible rise in access charges to the US delegation. After the talks, the representatives from the USTR spoke to the press about their serious concerns.

Since then, one question has been raised within the NCC circle. MIC had so far been offering a policy to support NCC by aiming at market competition and increasing new entries into the market. However, it was speculated that raising access charges did mean a shift in MIC's network policy. MIC was suspected of giving up its support of NCCs and shifting itself to passive, not active, protection of NTT.

Such speculation was made worse as an incident occurred: Professor Satoshi Daigo of Tokyo University failed to be reappointed as a member of the Telecommunications Council. It is customary for members of the Council to serve four terms, a totals of eight years, but MIC established a new rule to amend this eight-year term to six years and not to allow renewal. Professor Daigo did not get reappointed even though his status at the time made him eligible for reappointment. It is said that Professor Daigo gave opinions that did not necessarily favour MIC, and that there was a rift between them. Professor Daigo protested through the media about his removal, which he regarded as arbitrary in nature.

On 14 January 2003, MIC completed draft regulations to raise access charges by about 5 per cent and consulted the Council. The next day, the KDDI chairman, Onodera, protested at the press conference, saying, 'If access charges are raised, the competition in Japan's telecommunications market will be disturbed.' In the Japan–US talks on 27 February, the United States defended its stance, saying, 'The rise in access charges cannot be justified. This may slow down the NTT's restructuring process.'

Since this incident, the situation has largely changed. On 10 March, 18 new telephone operators submitted proposal documents concerning the access charge rise to MIC. However, despite the document submission, the Council adopted the decision to raise access charges on 28 March, while MIC promulgated the regulations regarding access charge amendments on 1 April. NTT East and NTT West applied to MIC for the access charge rise on 18 April and it was approved on 22 April.

Running parallel to this series of events in Japan was action by the

USTR: on 2 April, it announced an annual report based on Article 1377 of the Omnibus Trade and Competitiveness Act and criticized the Japanese access charge issue.[17] On 17 April, MIC submitted to the USTR a document protesting against these criticisms. On the same day, the KDDI chairman, Onodera, commented at a press conference that 'we are discussing all available best countermeasures, including legal actions', and this shocked the public, since his comment implied that NCC might be able to sue MIC, which was completely unprecedented. On 10 May, the executives of five telecommunications operators (KDDI, Japan Telecom, POWERDCOM, Cable & Wireless IDC and Fusion Communications) had a meeting and confirmed preparations for litigation. On 18 June, the KDDI chairman, Onodera, announced in public that they were in the process of preparation for the litigation. Responding to this move, MIC minister, Toranosuke Katayama, commented at the press conference after the cabinet meeting, saying, '[The Court] may dismiss the appeal.' However, these five executives again held a meeting on 10 July and decided to file administrative litigation. On 16 July, KDDI, POWERDCOM and Japan Telecom met Minister Katayama and delivered an official notice that they were filing an administrative litigation. Next day, five operators officially filed a lawsuit. In the second oral hearing, held on 28 November, the five operators submitted a test result of access charge burdens of FY 2003, which they estimated at US$33.7 billion in total.

On 27 February 2004, NTT East and West announced that the access charge of FY 2003 would be $33.4 billion. This amount represents 0.77 per cent of the operating earnings of FY 2003 of about $43.1 billion, and 18.0 per cent of ordinary income. Although the ratio within the operating earnings was small, the ratio in the ordinary income was large.

On 3 March, the study group of MIC finalized a new model for access charge calculation, and announced that 'it is possible to cut by a further 11.2 per cent'. On 20 April the Council's electronics and telecommunications study group started a discussion concerning a new model of calculation. Then the Council decided to adopt a policy of inhibiting the rise in access charges on 13 July. At the same time, however, it requested NTT to defer the basic charge that its customers were paying and to keep it at the current level. On 28 August, NTT East and West commented on the policy plan produced by the Council, protesting that 'absorption [of the loss without a rise in the basic charge] is impossible' in terms of their burdens in costs.

Since the middle of 2004, however, there has hardly been any media coverage about access charges for the fixed-line telephone system. At the time of writing, in December 2004, although the litigation case by NCC was in progress, new issues, which could make negotiation results less meaningful, emerged. It is the '*Otoku Line*' introduced by the Softbank Group and access charges for mobile communications (or mobile termination) that started drawing more attention.

Mobile phone access charge

The access charges incurred on mobile phones are a charge that the fixed-line telephone operators and international-call telephone operators pay to mobile phone operators. Normally, the line usage charge that customers pay is to be determined by an origin operator from which a call is made. However, in the Japanese mobile phone market, NTT DoCoMo established a custom when they launched the business, by which the receiving operator of a call (mobile operator) is allowed to determine the level of charge. Therefore, a mobile operator could determine the line usage charge in relation to a call coming from a fixed-line telephone to a mobile phone.

This is a very advantageous right for mobile operators. They can set charges whereby calls that are made from a mobile phone are cheaper than calls from a fixed phone, to make consumers use mobile phones more. Therefore, it was not mobile operators, but fixed-line operators that were able to levy the charges for the calls received, and the mobile phone operators could receive access charges without chasing those involved. In other words, the mobile phone operators could leave a stone unturned as far as they could maintain an advantageous position. Accordingly, some experts and the USTR raised concerns about the unique situation of mobile phone access charges. To avoid direct criticism, NTT DoCoMo announced on 25 March 2002 that it would lower the access charges by a maximum of 14.2 per cent. On 4 April, NTT DoCoMo's Keiji Tachikawa commented at the press conference that 'we are charging fairly [in relation to access charges]'.

As the rise in access charges of the fixed-line telephone system became apparent, the issue of mobile phone access charges was covered less. However, on 5 November 2002, the Telecommunications Business Dispute Settlement Commission within the MIC finalized a report that requested the transfer to the fixed-line telephone operators of a right to set the charges for a call coming from a fixed-line telephone to a mobile phone. NTT DoCoMo protested against this report but the MIC sustained its position regarding the right to set the charges and commenced the discussion on a new system on 30 January 2003, and the study group held hearings involving each operator. The issue was resolved on 25 June, when MIC officially announced the rendering of the right to set charges to the fixed-line telephone operators.

The actual services were commenced on 1 April 2004 and customers are entitled to a concession on the charge by dialling '0036' or '0039' before dialling normal mobile numbers.

The shock of 'Otoku Line'

What shone a spotlight on the issues related to access charges in the latter half of 2004 was Softbank Group's entry into the fixed-line telephone business.

Although less attention was initially paid to their moves, the trigger for this move was located in the lower setting of dry copper rental charges by NTT in autumn 2003. Dry copper is a copper cable that NTT owns in an area between customers' houses and its operational bureau offices. The price of dry copper was lowered to ¥1,366, which is cheaper than the basic line rental charge of ¥1,750 offered by NTT. As a result, it became a likely scenario that other operators can offer their services at a cheaper rate than the basic charge of NTT if they use their own switching devices, not NTT's device, situated inside NTT's bureau offices.

Softbank, by realizing this advantage, purchased Japan Telecom in July 2004 and suggested a new plan called '*Chokushu* [connecting directly] Service', which connects NTT's dry copper lines to NCC's own switching devices inside NTT's bureau offices. On 30 August, the Softbank chairman, Masayoshi Son, announced the commencement of a new fixed-line telephone service from December, the basic charge of which comes cheaper by around ¥200 than that of NTT.

This move by Softbank stimulated KDDI. It announced on 15 September that it would launch a new fixed-line telephone service cheaper than Softbank's by March 2005. Without missing an opportunity, Softbank announced on the same day a plan to offer further reduced-price services. Softbank pledged that it would provide the public with the cheapest services and of course cheaper than those of NTT (see Table 8.2).[18]

NTT was not quiet after Softbank's announcement, and it announced a lower basic charge on 1 October. It is ironic that NTT entered the price competition regarding basic charge against Softbank (Japan Telecom) and KDDI, because NTT had insisted on raising the basic charge if access charges were lowered.

Simultaneously, the abolition of a one-off connection fee[19] was being discussed. Japan Telecom and KDDI offered a free connection in their service package. Therefore, new customers were given a choice between

Table 8.2 Fixed telephone service prices (home use, without tax, yen)

	NTT Group (before price cut)	*NTT Group (after price cut)*	*KDDI*	*Softbank Group*
Monthly Basic Fee (with 'Push Phone' Function)	1,840–2,140	1,500–1,600	1,600 (flat rate)	1,450–1,600
Local (3 minutes)	8.5	7.5–8.5	8.0	7.9
Outside prefecture (3 minutes)	20–40	7.5–8.5	15 (flat rate)	14.9 (flat rate)
Inter-prefecture (3 minutes)	20–80	15 (flat rate)	15 (flat rate)	14.9 (flat rate)

Source: *Nihon Keizai Shinbun*, 1 January 2005, p. 39.

services offered by NTT and others; a customer could pick up a service with a one-off connection fee of ¥72,000 from NTT or a service with free connection from other operators. NTT felt that it was facing a crisis and suggested the gradual abolition of a one-off connection fee. On 12 October, a study group of the Telecommunications Council finalized a report on the approval of the abolition of a one-off connection and officially announced it on 19 October.

It was concerned, however, that if the abolition of a connection fee was announced, the existing customers might reject the decision. This is because a connection fee had been treated as some sort of property value to date and therefore its value would now become null and void, turned into just paper value.

The spread of '*Chokushu Service*' such as Otoku Line or KDDI's Metal Plus service is changing conditions in the fixed-line telephone market because it makes nominal the concept of access charge, which has been a prerequisite of the fixed-line telephone system. For NTT, which can levy access charges equivalent to the amount of dry copper, it is inevitable that a fall in revenues will be experienced. According to a test calculation by NTT, its revenue will fall by about US$13 billion, provided that it adjusts the charge to be equal to that of the cheapest operator.[20]

As a result, the spread of '*Chokushu Service*' implies the beginning of full competition in the local telecommunications market and is a sign that the issues of access charges are becoming less meaningful. The purpose of introducing access charges was to open facilities to others at a reasonable price, which had been regarded as a bottleneck. However, other operators do not find it necessary to negotiate with NTT if they do not have to rely on NTT's facilities and may be able to offer various forms of services.

This point of contention is applied to the same degree to foreign operators conducting business in Japan. While they may choose to connect their own networks to NTT's switching device as they did before, they have more choices to provide services at a cheaper rate by installing their own devices in NTT's operational bureau offices.

Conclusion

In this chapter, a political analysis was made of the issue of access charges that was put on the agenda during the Japan–US talks. Usually in goods and service trade negotiations, national interests can be clearly defined and it is a matter of how the negotiators pursue outcomes. However, in these access charge negotiations, Japan experienced extreme difficulties in defining its national interests. MIC, which failed to define them to the last minute, could not reach an agreement with the USTR. The reason why it failed is that the domestic actors deepened the conflicts by pursuing their own interests and this created a situation in which MIC did not have a chance to respond appropriately to the demands and criticisms from the

United States. In other words, since they could not agree at Level 2 in a two-level game, an agreement at Level 1 was also found difficult.

However, it is likely that the concept of access charges itself may become obsolete in the near future. For instance, mobile phone operators, who used to set access charges that were slightly different from those of the fixed-line telephone system, now face a new phase in which such a setting of access charges is determined by fixed-line telephone operators or foreign telecommunications operators.

Furthermore, with the arrival of '*Chokushu Service*', the NTT facilities needed have become merely dry copper and space in bureau offices for other operators. This in turn means falling revenue from NTT's access charges, and it is not beyond imagining that NTT may face a difficult business management situation in the future, including possible abolition of its one-off connection fee.

It seems that the Japan–US talks on access charges aimed only at the fixed-line network are not likely to occur any more. However, there is still room for discussion on access charges for the mobile phone sector.

To sum up, as long as there is a structural conflict between NTT and NCCs in Japan's domestic market, it is highly likely that the negotiations on access charges will continue in Japan. However, in any form of Japan–US negotiations in the future, it is important that Japan should precisely judge conditions it and its opposite numbers can accept.

MIC has two roles to play: it is a regulator of the market and a negotiator with a foreign counterpart. It seems that these two roles have produced a conflictual and difficult situation for MIC, which is otherwise supposed to force the competing operators to reach compromises, define national interests and take tough measures in negotiations without any obstacles. As a result, the Japan–US talks on the access charge became a difficult case to resolve for Japan.

Notes

1 James Rosenau, *Linkage Politics: Essay on the Convergence of National and International Systems* (New York: Free Press, 1969).

2 Karl W. Deutsch, Sidney A. Burrall, Robert A. Kann *et al.*, *Political Community in the North Atlantic Area: International Organization in the Light of Historical Experience* (Princeton, NJ: Princeton University Press, 1957); Ernst B. Haas, *The Uniting of Europe: Political, Social, and Economic Forces, 1950–1957* (Stanford, CA: Stanford University Press, 1958).

3 Robert O. Keohane and Joseph S. Nye, *Power and Interdependence*, 3rd edn (New York: Longman, 2001).

4 Graham T. Allison, *Essence of Decision: Explaining the Cuban Missile Crisis* (Boston: Little, Brown, 1971).

5 Peter J. Katzenstein (ed.) *Between Power and Plenty: Foreign Economic Policies of Advanced Industrial States* (Madison: University of Wisconsin Press, 1978); Stephen Krasner, *Defending the National Interest: Raw Materials Investments and U.S. Foreign Policy* (Princeton, NJ: Princeton University Press, 1978).

6 Robert D. Putnam, 'Diplomacy and Domestic Politics: The Logic of Two-Level Games', *International Organization* 42 (3): 427–460 (Summer 1988).

7 The term 'TELRIC' is used more in the United States.

8 Sogo Kenkyu Kaihatsu Kiko Denki Tsushin Sngyo Kenkyukai (Dai 2 Ki), *Denki Tsushin Sangyo ni okeru Jigyoshamou Sougo Setsuzoku ni kansuru Kenkyu* (Study on Operators' Interconnection in Telecommunications Industry) (Tokyo: Sogo Kenkyu Kaihatsu Kiko, 1998).

9 Takanori Ida, *Network Economics* (Tokyo: Nihon Hyoronsha, 2001, p. 78, footnote 7).

10 Motohiro Tsuchiya and Adam Thierer, 'Is America Exporting Misguided Telecommunications Policy?: The U.S.–Japan Telecom Trade Negotiations and Beyond', *CATO Institute Briefing Papers*, No. 79, 7 January 2002; Motohiro Tsuchiya and Atsuhiro Suzuki, 'Nichibei Keizai Masatsu no Henyo: NTT Setsuzokuryo no Seiji Bunseki' (Changes of Japan–U.S. Economic Frictions: Political Analysis of NTT's Interconnection Fee', *GLOCOM Review* 5(3): 1–25 (2000); Motohiro Tsuchiya, '1990 Nendai no Joho Tsushin Seisaku: NTT Keiei Keitai Mondai ni Torawareta Junen' (Communications Policy in 1990s: A Decade Captured by NTT Division Problem), *Leviathan* 32: 72–96 (2003); Tetsu Machida, *Kyodai Dokusen: NTT no Shukuzai* (Big Dominance: NTT's Original Sin) (Tokyo: Sinchosha, 2004).

11 The Ministry of Post and Telecommunications (MPT) was reorganized as the Ministry of Public Management, Home Affairs, Posts and Telecommunications (MPHPT) in 2001. Then the ministry changed its English name again in 2004. The new name is the Ministry of Internal Affairs and Communications (MIC).

12 However, access charges as discussed here have no direct impact on Internet expansion because Internet connection at that time was mainly dial-up (ADSL had prevailed since 2001). A dial-up user normally got connected through a local connection and therefore, the lowering of access charges would not have a huge impact.

13 Tokyo Electric Power Company has already established its own optic-fibre networks to measure gas usage of each household. By converting such a network to communications use, it proceeded to enter the communications market.

14 Although networks between houses and telephone operators' bureau offices can be borrowed, this service means the one connecting copper cables directly to systems run by other companies but situated inside NTT properties. Since this system makes minimal use of NTT's facilities, cheap services can be provided.

15 Japan's fiscal year starts in April and ends in March.

16 United States Trade Representative, 2002 Regulatory Reform Initiative Report, http://www.ustr.gov/assets/World_Regions/North_Asia/Japan/Regulatory_Reform_Initiative/asset_upload_file59_6533.pdf.

17 United States Trade Representative, 2003 Regulatory Reform Initiative Report, http://www.ustr.gov/assets/World_Regions/North_Asia/Japan/Regulatory_Reform_Initiative/asset_upload_file290_6530.pdf.

18 NTT Group's basic fee (after the price cut) does not include a paper bill.

19 It is usually called 'subscription right' (*Kanyu-ken*) in Japan. Its formal name is 'defrayment for facility installation' (*Shisetsu-secchi-futan-kin*). Everyone must pay it when they apply for a fixed-line telephone service. NTT and the former NTTPC used it to invest in rural areas under their universal service obligation.

20 NTT East, 'Comments on the existing method to calculate the connection fee as well as the impact of entry of *Chokushu Service* operators that use dry copper', p. 7 (28 September 2004).

Index